Modern Fisheries Engineering

Modern Fisheries Engineering

Realizing a Healthy and Sustainable Marine Ecosystem

Edited by
Stephen A. Bortone and Shinya Otake

CRC Press
Taylor & Francis Group
Boca Raton London New York

CRC Press is an imprint of the
Taylor & Francis Group, an **informa** business

First edition published 2020
by CRC Press
6000 Broken Sound Parkway NW, Suite 300, Boca Raton, FL 33487-2742

and by CRC Press
2 Park Square, Milton Park, Abingdon, Oxon, OX14 4RN

First issued in paperback 2022

© 2021 Taylor & Francis Group, LLC
CRC Press is an imprint of Taylor & Francis Group, an Informa business

Visit the Taylor & Francis Web site at
http://www.taylorandfrancis.com

and the CRC Press Web site at
http://www.crcpress.com

ISBN: 978-0-367-56066-9 (pbk)
ISBN: 978-0-367-34792-5 (hbk)
ISBN: 978-0-429-32803-9 (ebk)

DOI: 10.1201/9780429328039

Typeset in Times
by Deanta Global Publishing Services, Chennai, India

Contents

Preface

This book is largely based on presentations offered at the International Conference on Fisheries Engineering (ICFE) 2019: Realizing a Healthy Ecosystem and Sustainable Use of the Seas and Oceans. The conference was held in Nagasaki (Japan) on September 21–24, 2019 and included invited keynote addresses and general submissions centered on research in Fisheries Engineering. This larger topic included fisheries and environments associated with artificial reefs and seaweeds, construction of fishing ports, fishing boat and fishing gear engineering, resource survey technologies, and information technologies relating to fisheries, which contribute to mitigating global-scale issues such as climate change, food security and safety, and energy supply. The Japanese Society of Fisheries Engineering held this international conference to permit a review of today's Fisheries Engineering research and offer a platform for the discussion of future research directions. Another objective of the conference was encouraging future generations of young men and women in the field of Fisheries Engineering.

The conference site, Nagasaki, was an important location for this conference as the seas off Nagasaki have been designated as a Marine Renewable Energy demonstration field site by the Headquarters for Ocean Policy, Cabinet Secretariat in 2014. Consequently, research and development of related technologies are progressing in the region through collaborations with industries, administrative agencies, and academia. To further call attention to these accomplishments, this international conference also included a logical theme of "Marine Renewable Energy and Fisheries Engineering," and a one-day symposium on this topic.

Acknowledgments

The authors thank the presenters and attendees of the International Conference on Fisheries Engineering (ICFE) 2019. The presentations and fruitful discussions offered by all in attendance made the conference a success. Notably, the tangible evidence of these presentations is offered herein.

The conference (ICFE 2019) was organized by the Japanese Society of Fisheries Engineering and co-organized by Nagasaki University. We thank the many institutions that sponsored the conference: the Japan Fisheries Agency, Nagasaki Prefectural Government, Nagasaki City, Marinoforum 21, Japan Fisheries Cooperative Nagasaki, Japan Liaison Council for Fisheries and Marine Science Research, Fisheries Infrastructure Development Center, National Association of Fisheries Infrastructure (Japan), the Japanese Institute of Fisheries Infrastructure and Communities, All Japan Fishing Port Construction Association, Alpha Hydraulic Engineering Consultants Co., Ltd, AquaSound Inc. Asia Air Survey Co., Ltd, Association for Innovative Technology on Fishing Ports and Grounds, Environment Simulation Laboratory Co., Ltd, Japan Aqua Tec Co., Ltd, Kokusai Kogyo Col, Ltd, Lighthouse Inc., Nagasaki Fisheries Infrastructure Development Association, Nagasaki Ship Supplies Co., Ltd, Nippon Kaiyo Co., Ltd, Ocean Construction Co., Ltd, Okabe Co. Ltd, Senc21 Engineering Corporation, Toyo Corporation.

Lastly, we thank Alice Oven and her staff, especially Damanpreet Kaur, at CRC Press, as well as Bryan Moloney at Deanta Global, for their diligence and careful editing during the production of this volume.

Editor Bio

Stephen Bortone
Recently retired as Executive Director of the Gulf of Mexico Fishery Management Council, Stephen A. Bortone is now an Environmental Consultant with Osprey Aquatic Sciences, LLC based in Windham, New Hampshire (USA). As a consultant, he specializes in fisheries and is noted especially as an authority on artificial reefs. Currently, he is also the Marine Biology Series Editor with CRC Press. Previously, Dr Bortone served as Director of the Minnesota Sea Grant College Program, with an appointment as Professor of Biology at the University of Minnesota Duluth. Earlier in his career he was the founding Director of the Marine Laboratory at the Sanibel-Captiva Conservation Foundation in Sanibel, Florida, Director of Environmental Science at the Conservancy of Southwest Florida, and Director of the Institute for Coastal and Estuarine Research while Professor of Biology at the University of West Florida. Dr Bortone received the B.S. degree (Biology) from Albright College in Reading, Pennsylvania (1968); the M.S. degree (Biological Sciences) from Florida State University, Tallahassee (1970); and the Ph.D. (Marine Science) from the University of North Carolina, Chapel Hill (1973).

He has authored more than 180 scientific articles on the broadest aspects of the aquatic sciences over the past 50 years. In addition, he has edited seven books on the aquatic sciences: *Sea Grasses*; *Biology of the Spotted Sea Trout*; *Estuarine Indicators*; *Artificial Reefs in Fisheries Management*; *Interrelationships between Corals and Fisheries*; and *Red Snapper Biology in a Changing World* with S.T. Szedlmayer (all with CRC Press), in addition to *Marine Artificial Reef Research and Development: Integrating Fisheries Management Objectives* with the American Fisheries Society.

Internationally, he was a Visiting Scientist at the Johannes Gutenberg University (Mainz, Germany) and conducted underwater fish surveys with colleagues from La Laguna University in the Canary Islands for several years. He was also a Mary Ball Washington Scholar while conducting research and teaching at University College Dublin, Ireland. Notable awards and honors include: "Fellow" from the American Institute of Fishery Research Biologists, "Certified Fisheries Professional" from the American Fisheries Society, and "Certified Senior Ecologist" from the Ecological Society of America.

Shinya Otake
Dr Shinya Otake is a Japanese Fisheries Engineering researcher, and currently serves as Professor at the Marine Environmental Engineering Laboratory, Fukui Prefectural University. He is Chairman of the Design Committee for fisheries infrastructure projects implemented by the Japan Fisheries Agency, with specific regard to its fishery facilities. Dr Otake's research interests are on the development of artificial reefs and artificial upwelling. In addition, he was instrumental in incorporating acoustic engineering into the design of marine ranching in collaboration with the Penta Ocean Construction Co. Ltd from 1980 to 1992. He moved to Fukui Prefectural University in 1993 and obtained a doctorate from the University of Tokyo in 1997 while researching the development of upwelling structures. During the past 40 years, he has contributed numerous publications to the Japanese scientific literature. Since its origins in 1990, Dr Otake has been instrumental in the organization and promulgation of the Japanese Society of Fisheries Engineering. In recognition of his professional expertise in fisheries, he has served as the Society's Chairman for four terms since 2012. Since becoming Professor at the Fukui Prefectural University, he has also actively participated in CARAH (Conference of Artificial Reefs and Related Artificial Habitats, an international conference on artificial reefs), playing an active role in the conferences since the fourth meeting, held in 1987. He is also known for his active participation in most international artificial reef conferences, including participation in the RECIF conference on artificial reefs held in France in January 2015.

Contributors

Toru Aota
Association for Innovative Technology on
    Fishing Ports and Grounds
Tokyo, Japan

Stephen A. Bortone
Osprey Aquatic Sciences, LLC
Windham, New Hampshire

David De Monbrison
B.R.L Ingénierie Co.
Avenue Pierre Mendès, Nîmes, France

Kathryn H. Ford
Massachusetts Division of Marine Fisheries
New Bedford, Massachusetts

Yousuke Fukui
Association for Innovative Technology on
    Fishing Ports and Grounds
Tokyo, Japan

Oh Tae Geon
Korea Fisheries Resources Agency
Busan, Republic of Korea

Yasuyuki Gonda
Association for Innovative Technology on
    Fishing Ports and Grounds
Tokyo, Japan

Kengo Hashimato
Nagasaki branch
Fisheries Infrastructure Development Center
Nagasaki, Japan

Osamu Hashimoto
Association for Innovative Technology on
    Fishery Ports and Grounds
Tokyo, Japan

Baek Sang Ho
Korea Fisheries Resources Agency
Busan, Republic of Korea

Takeshi Hosozawa
Association for Innovative Technology on
    Fishing Ports and Grounds
Tokyo, Japan

Nariaki Inoue
Fisheries Technology Institute (Kamisu
    Branch)
Japan Fisheries Research and Education
    Agency
Kamisu, Japan

Satoshi Ishimaru
Nagasaki branch
Fisheries Infrastructure Development Center
Nagasaki, Japan

Syouichi Ito
Nagasaki branch
Fisheries Infrastructure Development
Center Nagasaki, Japan

Jarina Mohd Jani
Faculty of Science and Marine Environment
Universiti Malaysia Terengganu
Kuala Nerus, Terengganu, Malaysia

Minoru Kanaiwa
Graduate School of Bioresources
Mie University
Tsu City, Japan

Masaru Kawagoshi
Association for Innovative Technology on
    Fishing Ports and Grounds
Tokyo, Japan

Nobuo Kimura
Faculty of Fisheries Sciences
Hokkaido University
Hakodate, Hokkaido, Japan

Junji Kuwamoto
Nagasaki branch
Fisheries Infrastructure Development Center
 Nagasaki, Japan

Kim Jong Kyu
Chonnam National University
Jeonnam, Republic of Korea

Juliano Silva Lima
Centre of Bioscience and Biotechnology
University of North Rio de Janeiro, Campos
    dos Goytacazes
Rio de Janeiro, Brazil
and
Federal Institute of Education, Science and
    Technology
Nossa Senhora da Glória
Sergipe, Brazil

Motobumi Manabe
Association for Innovative Technology on
    Fishing Ports and Grounds
Tokyo, Japan

Takahito Masubuchi
Graduate School of Bioresources
Mie University
Tsu City, Japan

Jun Miyoshi
Fisheries Technology Institute (Kamisu
    Branch) Japan
Fisheries Research and Education Agency
Kamisu, Japan

Hideaki Nakata
Faculty of Fisheries
Nagasaki University
Nagasaki, Japan

Hirokazu Nishimura
Association for Innovative Technology on
    Fishing Ports and Grounds
Tokyo, Japan

Lee Moon Ock
Chonnam National University
Jeonnam, Republic of Korea

Fumihisa Okashige
Association for Innovative Technology on
    Fishing Ports and Grounds
Tokyo, Japan

Shinya Otake
Fukui Prefectural University
Gakuen-cho, Obama, Fukui, Japan

Sylvain Pioch
UMR 5175, Centre d'Ecologie Fonctionnelle et
    Evolutive
Université de Montpellier - Université Paul-
    Valéry Montpellier
Montpellier, France

Michael V. Pol
Massachusetts Division of Marine Fisheries
New Bedford, Massachusetts

Kimiyasu Saeki
Japan Fisheries Research and Education
    Agency (Kamisu Office)
Kamisu, Japan

Tetsuya Shirokoshi
Association for Innovative Technology on
    Fishing Ports and Grounds
Tokyo, Japan

François Simard
International Union for Conservation of Nature
    Commission on Ecosystem Management
Nice, France

Takeshi Tajima
Association for Innovative Technology on
    Fishing Ports and Grounds
Tokyo, Japan

Hideyuki Takahashi
Japan Fisheries Research and Education
    Agency (Kamisu Office)
Ibaraki, Japan

Tomomi Terajima
Association for Innovative Technology on
    Fishing Ports and Grounds
Tokyo, Japan

Akira Watanuki
Association for Innovative Technology on
    Fishing Ports and Grounds
Tokyo, Japan

Kenji Yasuda
Japan Fisheries Research and Education
    Agency (Kamisu Office)
Kamisu, Japan

Ilana Rosental Zalmon
Centre of Bioscience and Biotechnology
University of North Rio de Janeiro, Campos
    dos Goytacazes
Rio de Janeiro, Brazil

1 Introduction

Stephen A. Bortone and Shinya Otake

Fisheries Engineering is a field of study and investigation that likely had its origins when the earliest humans pursued fish as a resource. In its modern form, the discipline can cover the broadest range of topics, including: fish passageways, habitat protection and restoration, fishing gear development, bycatch reduction, hatchery management, artificial reefs, and aquaculture, among others. By way of example, the introduction to the Massachusetts Institute of Technology's Sea Grant program for Fisheries Engineering and Aquaculture has as its goals to investigate: fisheries, aquaculture, aquatic natural resources, food supply, employment, economic and cultural benefits from the aquatic environment, and investigations to support fishing communities, industries as well as commercial, recreational, and subsistence fisheries (MIT 2020). Similarly, the purpose of the Fisheries Engineering program in the Fisheries Sciences Department at the Hokkaido University has several goals. Among these are: maintain sustainable fisheries, rationally manage the marine environment and its fisheries resources, develop fishing vessels and fisheries research equipment, as well as analyze fisheries information and the behavior of living marine resources (Hokkaido University 2020).

One could easily argue that the ultimate goal of Fisheries Engineering is to sustain fisheries and its associated human socio-economic communities through the application and development of information and technology.

It is important to note that Fisheries Engineering has been interpreted as being responsible for the engineering sector of the fishery industry. In the past, fisheries (the fish and the associated human fishing community) were at the center of the fishing industry. Processing technology was needed to process the fish that had been caught. Engineering knowledge was required to streamline fishing and processing, and it was inevitable that Fisheries Engineering would be born there. Let us refer to these as the features of "old-school" Fisheries Engineering. Here let us consider what "new-school" Fisheries Engineering is. It starts with the self-awareness that the fishing industry is responsible for catching autonomous biological resources from the irreplaceable environment of the sea. Therefore, all aspects of the natural, social, and economic sciences are required to support the fishing industry. Fisheries Engineering is a conglomerate of these sciences, with artificial reefs as an associated environmental conservation technology.

Historically, Japan is one of the world's largest fisheries nations. Japan is dependent on the sea for its livelihood and way of life. In September 2019, an International Conference on Fisheries Engineering (ICFE2019) was held in Japan. The conference attracted more than 100 researchers from more than ten countries from around the world. The purpose of this international conference was to give an overview of the multi-faceted nature of Fisheries Engineering and to introduce new technologies for coastal environmental engineering. There was a special emphasis in the conference on addressing the future of fisheries and marine wind-power generation issues that are attracting attention around the world. Importantly, the conference also emphasized how artificial reefs are being incorporated into Fisheries Engineering.

This volume is important from several aspects, as it presents the achievements and advances of modern Fisheries Engineering by offering research results and perspectives on a broad range of topics. Because the conference at which these chapters were initially presented was held in Japan, the majority of chapters herein address issues in Japan. It should be noted that Fisheries Engineering is also included as part of fisheries science in various engineering departments of fisheries in Japan. Nevertheless, the general perspectives offered here have worldwide application.

A major focus of the ICFE2019 and the chapters here was the worldwide design and use of artificial reefs. Regarding artificial reefs, the Organizing and Steering Committees decided to include trends in artificial reefs around the world—including South America, Europe (chiefly France), Southeast Asia (mainly Malaysia), China, and Korea. In Japan, artificial reefs have been deployed since the Edo period (420 years ago), but many reefs are still being deployed today.

It is also important to realize that Fisheries Engineering in Japan differs somewhat from its comparable practice in the West, as in Japan, Fisheries Engineering includes environmental improvement engineering represented by artificial reef deployment and development, whereas in the West, artificial reefs are considered a facet of fisheries management. Based on the Japanese perspective, environmental improvement engineering (i.e., artificial reefs) is included here. Below is a brief introduction to the chapters in the volume.

Chapter 2 (Fisheries Engineering: Robust Fisheries for Today and Tomorrow by Nobuo Kimura) offers an introduction and review of the field of Fisheries Engineering from the perspective of the current research topics. This article introduces the keynote address of Dr Nobuo Kimura, who gave the purpose of the conference, and introduced the global trends of artificial reef initiatives. Moreover, this review of the literature, chiefly from the publication *Fisheries Engineering*, examines the development and application of fishery technology. This review also includes an examination of the potential impact of Fisheries Engineering on the future development of fishing vessels based on new regulations in Japan.

Chapter 3 (Trends and Obstacles in Artificial Reef Research by Juliano Silva Lima and Ilana Rosental Zalmon) examines the innovation that has occurred with the more widespread use of innovative materials and evaluation methods in artificial reef research. Notably, there are, however, persistent difficulties in conducting artificial reef research. These difficulties include the negative effects of some construction materials and continuing gaps in socio-economic data, as well as inequities in regional deployments of artificial reefs.

Chapter 4 (Artificial Reefs in France: Current State-of-the-Art and Recent Innovative Projects by Sylvain Pioch, David de Monbrison, and François Simard) offers an examination of past, present, and potential future use of artificial reefs in France to improve biodiversity and facilitate fisheries management. These structures, from the French perspective, have promise in helping to sustain fisheries and improve habitats through a balance of resource usage and thoughtful management.

Chapter 5 (Development and Utilization of Artificial Reefs in Korea by Lee Moon Ock, Oh Tae Geon, Baek Sang Hok, and Kim Jong Kyu) examines advances in artificial reef technology implementation through several innovative implementations to help sustain fisheries. These include seeding fish on artificial reefs and aligning artificial reefs with fish ranching activities. Evidence indicates these efforts have been successful thus far, with more success forthcoming through improved cooperation between local governments and fishing communities to improve environmental conditions.

Chapter 6 (The Status of Artisanal Fish Aggregating Devices in Southeast Asia by Jarina Mohd Jani) focuses on the attraction features of artificial reefs through the development of new materials and designs recently implemented by artisanal fishers. While there are significant problems with some designs and materials, these are being overcome through an integrated management approach and attention to regional/cultural features of Southeast Asia.

Chapter 7 (Design and Creation of Fishing Grounds in Japan with Artificial Reefs by Shinya Otake) introduces design guidelines for implementation of artificial reefs in Japan. These guidelines, based on previous efforts, facilitate the application of artificial reefs to improve and create fishing grounds. Future applications of artificial reefs made toward the improvement of fishing will benefit from input from fishermen and serve to guide future efforts in fisheries enhancement.

Chapter 8 (Using Standardized CPUE to Estimate the Effect of Artificial Reefs on Fish Abundance by Nariaki Inoue, Satoshi Ishimaru, Kengo Hashimato, Junji Kuwamoto, Takahito Masubuchi, and Minoru Kanaiwa) significantly elevates our ability to more accurately portray changes in fisheries when they have been enhanced by artificial reefs as a fishing habitat. This improvement in analysis has broad applications toward evaluating the impact of artificial reefs in habitats around the world.

Chapter 9 (Using Artificial Substrata to Recover from the Isoyake Condition of Seaweed Beds off Japan by Osamu Hashimoto, Motobumi Manabe, Akira Watanuki, Masaru Kawagoshi, Takeshi Hosozawa, Fumihisa Okashige, Yasuyuki Gonda, Syouichi Ito, Takeshi Tajima, Yousuke Fukui, Tomomi Terajima, Hirokazu Nishimura, Tetsuya Shirokoshi, and Toru Aota) examines efforts to overcome the impacts of isoyake (barren substrate) events off Japan. The devastating effect of these events has impeded the sustainability of many local coastal fisheries. The efforts outlined here include defenses against herbivores and improvements to the environment that include planting, fertilization, and deploying alternative substrates. Success depends on a balance between both structural and non-structural countermeasures.

Chapter 10 (Why do Japanese Fishermen Not Wear Life Jackets? Answers Based on Interviews with Fishermen by Hideyuki Takahashi, Kenji Yasuda, and Kimiyasu Saeki) takes up an examination of safety issues experienced by fishermen. Here, different safety vests were worn under normal working conditions. The conclusions that safety vest design must accommodate actual working situations will go far in giving direction to future efforts to refine this essential lifesaving equipment.

Chapter 11 (Habitat-Creation in the Sustainable Development of Marine Renewable Energy by Hideaki Nakata) considers the habitat creation that occurs when structures are deployed to develop renewable energy. The ever-increasing deployments of artificial habitats through the desirable creation of renewable energy sources has far-reaching implications for the future of fisheries habitat improvement. The application of artificial reef technology toward an alternative purpose can have a dual role in the betterment of our fisheries.

Chapter 12 (Offshore Wind Energy and the Fishing Industry in the Northeastern USA by Michael V. Pol and Kathryn H. Ford) continues with the efforts to create habitats from energy-based constructions, but through an examination of the political/regulatory aspects that confront habitat deployments. Here the case is made for the inclusion of a broad range of interested user-groups in the decision, development, deployment, and evaluation aspects of the effects of adding structures to the marine environment. While not strictly an artificial reef, the thought (and legal) processes of artificial reef deployments are similar to any actions that affect habitats (either purposeful or incidental).

Chapter 13 (Hydrogen Fuel Cell and Battery Hybrid Powered Fishing Vessels: Utilization of Marine Renewable Energy for Fisheries by Jun Miyoshi) expands the breadth of Fisheries Engineering by taking a modern, energy-saving examination of the conversion of a typical diesel-powered fishing vessel and offers a plan to convert its power source to a more energy-efficient, fuel cell design. This chapter demonstrates the future of much of our future efforts in Fisheries Engineering.

Chapter 14 (Summary: The Future of Fisheries Engineering by Stephen A. Bortone and Shinya Otake) presents an overview of practical and likely future avenues for investigations in Fisheries Engineering. The overview includes an expansion of the presentations offered here with a further discussion of near future needs and demands of the discipline. Areas for future research are also offered to, hopefully, serve as a guidepost for future generations of fisheries engineers.

Future compilations of Fisheries Engineering research will have been well served by the perspectives offered here. We sincerely hope that the summaries, guidance, and future directions posited will serve as serve as a firm platform for generations of fisheries engineers.

REFERENCES

Hokkaido University. 2020. *Hokkaido University Fisheries Sciences.* http://www2.fish.hokudai.ac.jp/language-english/graduate-school/fisheries-engineering/.

MIT. 2020. *Massachusetts Institute of Technology Sea Grant College Program.* https://seagrant.mit.edu/sustainable-fisheries-aquaculture/.

2 Fisheries Engineering
Robust Fisheries for Today and Tomorrow

Nobuo Kimura

CONTENTS

ABSTRACT

The many studies and research conducted in the field of Fisheries Engineering have significantly contributed to the modernization and development of fisheries and aquaculture. This review is based chiefly on the studies published in *Fisheries Engineering* by the Japanese Society of Fisheries Engineering. The development and application of fishery technology examined here includes four major fields of study: artificial reefs, underwater surveys, utilization of artificial intelligence, and the application of computational fluid dynamics technology. In addition, this paper examines the potential for technology in the future development of fishing vessels based on the Japanese fisheries law revised in 2018.

INTRODUCTION

Fisheries Engineering has played a crucial role in supporting fisheries, resulting in increased production, resource conservation, and operational changes mandated by global climate change. For many years, the ocean was viewed as a limitless resource. Today, however, the oceans face major threats including: climate change, pollution, habitat destruction, invasive species, and overfishing. Fisheries Engineering has benefited human society by addressing important issues affecting the sustainability of fisheries using the latest technology and extensive knowledge gained through field experiments.

In this chapter, a discussion is offered as to how Fisheries Engineering has contributed to fisheries and human society through the experience of extensive field studies.

CURRENT STATUS SURROUNDING FISHERY AND RESOURCES

Global capture-fishery production has remained relatively static since the late 1980s (FAO 2019). The FAO (Food and Agricultural Organization of the United Nations) estimated that from 1975 to 2015, the percentage of overfished stocks (i.e., stocks fished at biologically unsustainable levels) increased from 10% to 33%. At the same time, the percentage of underfished stocks (i.e., stocks with a potential for expansion) decreased from 39% to 7%. While capture-fishery production appears to have achieved its maximum, aquaculture production has increased dramatically, attaining 80 million tons in 2016 and accounting for 47% of global fish production. Aquaculture production is expected to continue growing to meet the worldwide demand for fish (FAO 2018).

Fishery production in Japan, which was once the world's leading country for fisheries landings, is now less than half of its maximum level of 1984 (MAFF 2018, Government of Japan 2017, 2018). In addition, Japanese fishers are decreasing in number with a concomitant increase in their average age. The Japan Fisheries Agency predicts that during the next 30 years, the number of fishers will decrease to about 70,000, which is about half the current level. To maintain sustainable fisheries and aquaculture under these conditions, Fisheries Engineering will need to develop and expand the use of advanced technologies such ICT (Information and Communication Technology), IoT (Internet of Things), and AI (Artificial Intelligence), etc.

SUSTAINABLE USE OF FISHERIES RESOURCES

Fisheries Engineering is currently addressing many fisheries-related problems by focusing on sustainable and efficient use of marine resources, as well as improving resource conservation. The critical problems caused by the vulnerability of the earth's ecosystem, coupled with increasing human activity, continue as challenges we now face as humans (Miura 2008). The following products will provide guidelines for the future of fisheries and Fisheries Engineering:

- Sustainable Development Goals (SDGs), UN (UN 2015)
- Society 5.0, Cabinet office, Japan (Government of Japan 2016)
- Smart fishery, Future Investment conference, Japan (Government of Japan 2017, 2018)
- Revised Fisheries Act, Japan (MAFF 2018a)

With these guidelines in mind, studies and research will be able to contribute to the development of sustainable and robust fisheries and aquaculture.

DEVELOPMENTS AND APPLICATION OF TECHNOLOGIES

The Japanese Society of Fisheries Engineering has held symposiums twice a year to discuss new issues being faced. The themes discussed in the symposiums for the last 5 years are as follows:

- Marine renewable energy and Fisheries Engineering, September 23, 2019
- Exploring Fisheries Engineering innovations that can respond to climate change, May 20, 2019
- Fisheries Engineering and artificial intelligence, October 6, 2018
- Current status of ICT utilization in the fishing port and fisheries professions, as well as problems associated with technological developments and their introduction, May 14, 2018
- Current status and technical issues formulating a vision about seaweed beds and tidal flats, December 9, 2017
- Current status and problems of counting and measuring technologies in fisheries and aquaculture, May 29, 2017
- Exploring the potential of Fisheries Engineering studies using drones, September 17, 2016

- Current status and issues of training ships required for fisheries education and research, June 4, 2016
- Comparison and consideration of purse seine fisheries in Japan and Norway from the viewpoint of systems, distribution, processing, and fishing boat operations and design, November 14, 2015
- Discussion on the rationalization of night-time squid fishing, May 31, 2015

Additionally, there has been a focus on using new technologies, such as artificial intelligence, the Internet of Things, drones, etc. Thus, Fisheries Engineering has always worked on solving problems associated with new issues about fisheries and aquaculture.

Below are highlighted five important research areas based on recent studies published in the journal *Fisheries Engineering*:

ARTIFICIAL REEFS

To enhance coastal fisheries, many artificial reefs have been deployed to improve productivity. The development of offshore fishing grounds has progressed using artificial mound-reefs (i.e., large-scale, high-profile artificial reefs) in recent years. There have been many artificial reefs deployed on the seabed, from nearshore coastal areas out to the deep sea. Many studies have examined how these structures attract and help propagate fishes. For example, Ito (2011) and Ito et al. (2018) summarized the current conditions in, and the creation of, fishing grounds with artificial reefs and determined the residence time and swimming depth of threeline grunt (*Parapristipoma trilineatum*) at an artificial mound-reef. Koike and Otake (2017) examined the effect of flow around different arrangements of artificial reefs. In response to a growing interest in biodiversity, Anaguchi et al. (2014) summarized studies about the biodiversity associated with artificial reefs. Komeyama et al. (2014), using acoustic telemetry near an artificial reef, investigated the effect of tidal currents on hatchery-reared red seabream (*Pagrus* spp.). These and other studies indicate that the use of various artificial reefs will continue to improve the productivity of fisheries resources in the future.

UNDERWATER SURVEYS

Accurate and precise information about fishes and the aquatic environment are indispensable for fisheries and aquatic resource management. To collect such information, acoustic instruments operated underwater are among the most common and effective. Fish-finders have been used since the 1950s. There are many investigations that have conducted acoustic studies to collect this information. For example, Takao (2012) traced the development and practical application of fisheries acoustics. Additionally, Hamano (2019) conducted many acoustic studies to estimate fishery resources and the extent of reefs on fishing grounds.

Recently, the need has arisen for new aquaculture management strategies that will help maintain and attain fishery production goals while reducing labor costs. Accurate information concerning cultured fish such as body length, body weight, and the optimal number of fish retained in cages or tanks is important for aquaculture management. Torisawa et al. (2012) proposed a three-dimensional monitoring method for cultured fishes, and Takagi et al. (2018) developed a non-contact measuring system for aquaculture production management using image analysis.

ARTIFICIAL INTELLIGENCE IN FISHERIES AND AQUACULTURE

Artificial intelligence deals with the simulation of intelligent thinking using computers, which allows complicated tasks to be performed flexibly and accurately. In Fisheries Engineering, the application of artificial intelligence technology with intense learning has been used for various difficult tasks. Ito (2019) constructed detailed bathymetric charts using image resolution via machine-learning.

Hashimoto et al. (2019) studied the use of artificial intelligence technology to prevent fishing vessel collisions. Arai and Dehara (2019) predicted potential fishing grounds from visible images of satellite data via machine-learning. To grasp the growth situation of coral, Ukai and Nakase (2019) used artificial intelligence for physical environments such as geographical features. Currently, artificial intelligence is being used in the construction of aquaculture systems to predict the occurrence of serious fish diseases (Rahman and Tasnim 2014, Rao et al. 2017). Artificial intelligence technology is expected to spread rapidly in various aspects of Fisheries Engineering (Mizukami 2019).

APPLICATION OF COMPUTATIONAL FLUID DYNAMICS TECHNOLOGY

Computer-aided engineering has been used in a wide range of industries and is now a fundamental technology in every stage of design, improvement, and/or development in fishery production planning. Recently, computational fluid dynamics (CFD) technology has rapidly spread in fisheries because it can effectively address physical problems associated with fluids such as seawater. In aquaculture, mass mortality is a serious problem for fishery managers, especially with regard to larval and juvenile life stages. Complex flow distribution in tanks has a significant impact on fishes without swimming capability in these stages, so it is important to understand and manage the water flow in holding tanks. Kotake et al. (2017) and Takakuwa et al. (2018) conducted flow simulations in tanks using computational fluid dynamics and suggested using the results of this approach to design aquaculture systems.

REVISED FISHERY LAWS IN JAPAN

For the first time in 70 years, Japan's Fisheries Act was revised in December 2018 by the legislative National Diet of Japan. The focus of the original Fisheries Act was to enhance the productivity of Japanese fisheries, while the revised Act also addressed new issues, such as sustainability of fisheries and the safety of fishers. Serious fishing accidents such as capsizing have led to the loss of many lives. Regarding the seaworthiness of fishing vessels and fisher safety aboard Japanese fishing boats (Amagai et al. 2000, Hamaguchi et al. 1995, 1995a, 2000, Kimura et al. 1999, 2013), the safety of fishers can be increased through restricting the gross tonnage of fishing boats (Kimura 2011, Uchida 2012, Miyoshi 2016). Regardless of the gross tonnage, however, it is possible to redesign fishing boats with higher efficiency, comfort, and safety under this new fishery policy. The new fishing policy has the potential to improve fishing boats and the lives of fishers in the future.

CONCLUSION

The Japanese Society of Fisheries Engineering holds biannual symposia to address new issues and conduct comprehensive discussions among a wide range of scientists. New problems will continue to arise, such as climate change and its effects on the ocean, which will in turn affect capture-fisheries and aquaculture. To deal with complicated issues such as these, fisheries engineers and scientists must continue to accurately assess the current state of Japanese fisheries. Additionally, efforts are needed to clearly define future goals and advance studies about Fisheries Engineering using the latest technology, with consistent measures such as the backcasting approach (Miura 2008). It is hoped that Fisheries Engineering will contribute to enhancing the future resilience and sustainability of fisheries and the fishing industry in Japan.

ACKNOWLEDGMENTS

To investigate studies about the Fisheries Engineering, I have referred to many papers published in *Fisheries Engineering*. I express my sincere appreciation for the authors of these studies.

REFERENCES

Amagai, K., K. Ueno, and N. Kimura. 2000. Characteristics of roll motion for small fishing boats. In: *Contemporary Ideas on Ship Stability*, eds. D. Vassalos et al. Elsevier Science, 137–148.

Anaguchi, Y., K. Nagamatsu, M. Tahara, and Y. Adachi. 2014. Some examples to study about the biodiversity in artificial reefs. *Fisheries Engineering* 50(3):219–224.

Arai, Y., and M. Dehara. 2019. Prediction the potential fishing grounds using machine learning and satellite data. *Fisheries Engineering* 56(1):57–60.

FAO. 2018. *The State of World Fisheries and Aquaculture 2018*. Food and Agriculture Organization of the United Nations. www.fao.org/3/I9540EN/i9540en.pdf (accessed January 9, 2020).

FAO. 2019. *Fishstat*. Food and Agriculture Organization of the United Nations. http://www.fao.org/fishery/statistics/software/fishstat/en (accessed January 9, 2020).

Government of Japan. 2016. *The 5th Science and Technology Basic Plan*. The Cabinet Office. https://www8.cao.go.jp/cstp/kihonkeikaku/5honbun.pdf (accessed January 9, 2020).

Government of Japan. 2017. *Future Investment Strategy 2017*. Future Investment Conference. http://www.kantei.go.jp/jp/singi/keizaisaisei/pdf/miraitousi2017_t.pdf (accessed January 9, 2020).

Government of Japan. 2018. *Future Investment Strategy 2018*. Future Investment Conference. http://www.kantei.go.jp/jp/singi/keizaisaisei/pdf/miraitousi2018_zentai.pdf (accessed January 9, 2020).

Hamaguchi, M., S. Shimokawa, N. Kimura, and K. Amagai. 1995. On the motions and the operational limitations of a purse seiner (135 GT size) searching for a shoal of fish. *Fisheries Engineering* 31(3):161–168.

Hamaguchi, M., N. Kimura, and K. Amagai. 1995a. On the motions and the safety of a 135 GT purse seiner during a cycle of shooting and hauling work. *Fisheries Engineering* 32(2):79–87.

Hamaguchi, M., N. Kimura, T. Hokimoto, J. Kawasaki, and K. Amagai. 2000. A study of the purse wire tension working on a 135 GT purse seiner during a fishing operation, -Estimate of tension working on the wire using a neural network model. *Fisheries Engineering* 36(3):243–248.

Hamano, A. 2019. Studies of acoustic method for estimating fishery resource and fishing reef ground. *Fisheries Engineering* 55(3):175–185.

Hashimoto, H., S. Haiqing, A. Matsuda, and Y. Taniguchi. 2019. Utilization of AI technology for collision avoidance of fishing vessels. *Fisheries Engineering* 56(1):51–55.

Ito, K. 2019. Efficient bathymetry by learning-based image super resolution. *Fisheries Engineering* 56(1):47–50.

Ito, Y. 2011. Creation of fishing ground・artificial reef and present condition. *Fisheries Engineering* 48(2):157–160.

Ito, Y., K. Nakamura, and T. Yoshida. 2018. Residence time and swimming depth of Threeline grunt *Parapristipoma trilineatum*, gathering around an artificial mound reef. *Fisheries Engineering* 55(1):21–27.

Kimura, N. 2011. Renewal of fishing boats and control of displacement. *Fisheries Engineering* 48(2):151–156.

Kimura, N., K. Amagai, T. Kodama, and T. Hokimoto. 1999. On the effectiveness of the forecasting for the rolling motion of a small fishing vessel using a neural network model. *Fisheries Engineering* 35(3):229–234.

Kimura, N., Y. Fujimori, H. Yasuma, and K. Maekawa. 2013. Estimation of dynamic frequency and time characteristics in ship motions of an offshore trawler. *Fisheries Engineering* 49(3):167–175.

Koike, S., and S. Otake. 2017. The effect to the flow around the artificial reef in the different arrangement in situ. *Fisheries Engineering* 53(3):139–147.

Komeyama, K., R. Takayanagi, K. Fujioka, Y. Yamanaka, T. Ojiro, and K. Anraku. 2014. The effect of tidal current on the movement of hatchery-reared read sea bream near fish aggregating devices. *Fisheries Engineering* 51(1):21–31.

Kotake, G., Y. Takahashi, H. Yasuma, K. Maekawa, and N. Kimura. 2017. Visualization of water flow in aquaculture tank using computational fluid dynamics analysis. *Fisheries Engineering* 54(2):97–105.

MAFF. 2018. *The Census of Fisheries*. The Ministry of Agriculture, Forestry and Fisheries, Japan. https://www.maff.go.jp/j/tokei/census/fc/2018/2018fc.html (accessed January 9, 2020).

MAFF. 2018a. *Revision of Fisheries Law*. The House of Representatives, Japan. The Ministry of Agriculture, Forestry, and Fisheries, Japan. http://www.shugiin.go.jp/internet/itdb_gian.nsf/html/gian/honbun/houan/g15109076.htm (accessed January 9, 2020).

MAFF. 2019. *Fisheries and Aquaculture Production Statistics*. The Ministry of Agriculture, Forestry and Fisheries, Japan. https://www.maff.go.jp/j/tokei/kouhyou/kaimen_gyosei/index.html (accessed January 9, 2020).

MAFF. 2019a. *FY2016 Trends in Fisheries FY2017 Fisheries Policy White Paper on Fisheries: Summary.* The Ministry of Agriculture, Forestry and Fisheries, Japan. https://www.maff.go.jp/e/data/publish/at tach/pdf/index-66.pdf (accessed January 9, 2020).

MAFF. 2019b. *FY2017 Trends in Fisheries, FY2018 Fisheries Policy White Paper on Fisheries: Summary.* The Ministry of Agriculture, Forestry and Fisheries, Japan. https://www.maff.go.jp/e/data/publish/at tach/pdf/index-94.pdf (accessed January 9, 2020).

MAFF. 2019c. *FY2018 Trends in Fisheries, FY2019 Fisheries Policy White Paper on Fisheries: Summary.* The Ministry of Agriculture, Forestry and Fisheries, Japan. https://www.maff.go.jp/e/data/publish/at tach/pdf/index-166.pdf (accessed January 9, 2020).

Miura, T. 2008. *Zero Emissions Approach to Sustainable Fisheries Science.* Hokkaido University: Hokkaido Univ. Press, 1–167.

Miyoshi, J. 2016. Comparison of basic design concept for Japanese and Norwegian purse seiner. *Fisheries Engineering* 53(1):37–50.

Mizukami, K. 2019. Application examples of artificial intelligence for effective inspection of coastal embankments. *Fisheries Engineering* 56(1):61–64.

Rahman, A., and S. Tasnim. 2014. Application of machine learning techniques in aquaculture. *International Journal of Computer Trends and Technology* 10(4):214–215.

Rao, Ventateswara P., A. Ramamohan Reddy, and V. Sucharita. 2017. Computer aided shrimp disease diagnosis in aquaculture. *International Journal for Research in Applied Science and Engineering Technology* 5II:538–541.

Takagi, T., K. Komeyama, S. Abe, et al. 2018. Development of non-contact measuring system for aquaculture using image analysis. *Fisheries Engineering* 54(3):209–213.

Takakuwa, Y., W. Yamazaki, T. Sumida, and Y. Sakakura. 2018. Flow field investigation in rectangular tanks by bubby flow simulations. *Fisheries Engineering* 54(3):155–162.

Takao, Y. 2012. Development and practical application of fisheries acoustics in these 20 years. *Fisheries Engineering* 48(3):231–233.

Torisawa, S., M. Kadota, K. Komeyama, and T. Takagi. 2012. A technique of three-dimensional monitoring for free-swimming Pacific Bluefin Tuna *Thunnus orientalis* cultured in a net cage using a digital stereo-video camera system. *Fisheries Engineering* 49(1):13–20.

Ukai, A., and K. Nakase. 2019. Utilization of physical environment information and AI in estimation of distribution of coral coverage. *Fisheries Engineering* 56(1):67–70.

UN. 2015. *Sustainable Development Goals, 2030 Agenda.* United Nations. https://sustainabledevelopment .un.org/.

Uchida, K. 2012. Study of fishing boats in fisheries engineering. *Fisheries Engineering* 48(3):235–242.

3 Trends and Obstacles in Artificial Reef Research

Juliano Silva Lima and Ilana Rosental Zalmon

CONTENTS

ABSTRACT

This study provides a review of the main published studies on artificial reefs to evaluate scientific trends, impacts and challenges for the development of artificial reef research. The review includes more than 620 studies published worldwide from 1962 to 2018. In general, publications on artificial reefs have focused on investigating the ability for artificial reefs to attract fishing resources and on constructing artificial reef structures with new materials. Artificial reef science has made significant advances in recent decades with the increased use of more elaborate construction methods and data analyses. Nevertheless, the field has encountered difficulties in minimizing the negative effects of some materials, filling gaps in socio-economic data, developing integrated management assessments and overcoming regional inequalities. There are promising opportunities to overcome these obstacles that depend on the development of inert materials, an increase in the number of long-term and large-scale studies, the regulation of an evaluation protocol, the development of conservation policies and investments in scientific dissemination. We hope that this systematic review can not only advance artificial reef science but also provide important information for overcoming obstacles to the future development of this topic.

INTRODUCTION

Artificial reefs are widely used for habitat restoration due to the need to increase biomass and the abundance of biological resources (Leitão 2013, Streich et al. 2017). These reefs can be considered to have a primary function when the installed structures have the objective of imitating some aspect of the natural environment. They have a secondary function when the structures are not primarily designed to attract biota, but rather for other purposes (i.e., oil/gas platforms, port structures, ships and tanks, and aquatic foundations for renewable energy systems), but function as artificial reefs once deployed.

The historical use of primary artificial reefs has included materials that are installed to provide new substrates for species settlement, feeding, breeding, and predator refuge (Taylor et al. 2018, Bull and Love 2019). Since the 1960s, the deployment of artificial reefs has increased, becoming a worldwide practice, especially in marine ecosystems in Japan, the USA, and Western Europe (Lima et al. 2019a). Despite their long empirical use by fishermen, the increased use of these structures has been related to fishery management, recreational diving, aquaculture, coastal erosion control, and conservation of biodiversity (Dagorn et al. 2013a,b, Chapman et al. 2018).

Artificial reefs are important elements in the coastal management plans of several countries (Guan et al. 2016, Bucaram et al. 2018). Most research on artificial reefs, to date, has focused on marine environments in subtropical regions (Firth et al. 2016, Bull and Love 2019, Cresson et al. 2019). Moreover, a considerable portion of the information currently available from scientific articles on artificial reefs has been concerned with investigations on the ecological aspects and behavioral issues associated with biological response to the engineering of new materials (Huang et al. 2016, Mohamad et al. 2016, Smith et al. 2017). Although great advances have been made in artificial reef science, many questions regarding the performance, management, socio-economics, and impacts of these structures remain unanswered (Becker et al. 2018, Lee et al. 2018, Lima et al. 2019a).

In the ecological context, significant advances have been made regarding the role of artificial reefs in species attraction-aggregation (Bohnsack 1989, Smith et al. 2015) and/or production (Lindberg 1997, Brickhill et al. 2005). In the socio-economic context, the creation of artificial habitats has emerged as part of an alternative management strategy to increase productivity and income from aquatic systems (Chen et al. 2013, Westerberg et al. 2013). Activities with this strategy include creating industrial fishing exclusion areas (Murillas-Maza et al. 2013) that benefit fishing communities (Macusi et al. 2017, Lima et al. 2019b).

Given the increase in the number of studies on artificial reefs in recent years and the need to better understand the challenges of research on these structures, a systematic review was conducted that included a characterization of these studies and an examination of the current challenges in artificial reef science. For practical purposes, this review was divided into three sections. The first section is the preparation of profiles of studies on artificial reefs that include an inventory of the main articles available in the major online academic databases published between 1962 and 2018. Second, there is an examination of obstacles currently confronting research on artificial reefs, describing the main problems and impacts generated by the use of artificial structures and how these factors limit the progress of research on the subject. Last, there is an exploration of how to overcome challenges regarding artificial reef issues, providing new perspectives for research on artificial habitats. It is expected that, in addition to offering a profile of the scientific landscape on the use of artificial structures, this study may also provide important information for overcoming obstacles for future studies on artificial reefs.

MATERIALS AND METHODS

Artificial reefs here were considered to be any structure installed in an aquatic habitat that acts as a substrate or shelter for marine resources. Primary reefs included fish aggregating devices (FADs) and benthic structures usually made of concrete, metal, rock, rubber, plastic, wood, rope, netting, clay, fiberglass, and geotextile. Port-associated structures (i.e., sea walls, rip rap, docks, piers, etc.), wind energy foundations, wave energy devices, shipwrecks, and oil and natural gas platforms were considered secondary reefs. Based on this definition, a systematic review (PRISMA method, following Moher et al. 2009) of the scientific literature on artificial reefs published from January 1962 to December 2018 was conducted.

A structured search was performed using four databases (Scopus®, Web of Science®, Scielo®, and Google Scholar®) and by searching the terms "artificial reef," "artificial structure," "artificial habitat," "artificial sea-mount," "surf reef," "fish aggregating device," "fish attracting device," "wind farm," and "wind facilities" in the title, abstract, or keywords of the articles. Although it is recognized that there are many studies published in the grey literature (i.e., non-peer-reviewed), information from these was not included in this review because the sources are generally difficult to access and rarely undergo a peer-review process.

The systematic search yielded 6,155 documents, of which 620 articles were selected for the review conducted here, based on four exclusion criteria. The first criterion excluded books, theses, dissertations, and technical reports in the online databases (2,935 documents were excluded by this criterion). The second criterion excluded duplicate articles in the databases (1,269 documents were excluded by this criterion). The third criterion excluded articles for which the full text was not available online (806 documents were excluded by this criterion), and finally, articles whose central theme was not artificial reefs were excluded (525 documents were excluded by this criterion).

The following metadata were recorded from the selected articles: 1) year of publication, 2) country of the institution of the first author, 3) study objective, 4) location of the study area (country and geographical coordinates), 5) type of artificial reef (primary or secondary reef), 6) sampling methodology, and 7) main results and conclusions of the studies.

RESULTS AND DISCUSSION

PROFILES OF STUDIES ON ARTIFICIAL REEFS

The articles reviewed describe the different focuses related to artificial reefs and indicate the increased global interest in the objectives of these structures. Of the 620 articles reviewed, 314 addressed biological or ecological characters of fish, invertebrates, macroalgae, and plankton. Significant theoretical advances were also reported in the fields of engineering (n = 67), socio-economics (n = 49), management (n = 41), animal behavior (n = 38), environmental impacts (n = 29), and oceanography (n = 27). Most studies (n = 452, 73%) were conducted on artificial reefs designed primarily to attract fishing resources (Figure 3.1).

The primary reefs examined here were constructed of concrete (n = 289), metal (n = 147), rock (n = 82), FADs (n = 45), PVC (n = 27), and biogenic material (n = 23). The studies examined here on secondary reefs were, structurally, oil/gas platforms (n = 83), ships and shipwrecks (n = 49), port structures (n = 37), and renewable energy plants (n = 15). The different objectives for deploying primary and secondary artificial reefs varied with regard to ecological effects, the intended species for attraction, and the implications of these structures on the environment (Seaman Jr. 2000, Brickhill et al. 2005, Evans et al. 2017).

The efforts to develop artificial reef science has differed over time and among countries. Fifty-six countries were ascribed to studies on artificial reefs (Figure 3.1). The USA (n = 155), China (n = 60), Australia (n = 51), Brazil (n = 48), Italy (n = 38), and the United Kingdom (n = 37) were the main countries that investigated this subject (Figure 3.1). Other countries include Japan (n = 32), France (n = 30), Portugal (n = 28), Spain (n = 21), Israel (n = 18), and South Korea (n = 13). The scientific output of scientists from these countries corresponds to 84.9% of the publications. Moreover, the differences in the research topics according to country may be related to the environmental management activities designed to resolve different regional problems (Figure 3.2).

The deployment of artificial reefs has helped direct research that is focused on manipulation/experimentation to clarify settlement and ecological succession processes; however, a large number of the studies examined here were exclusively observational (Lee et al. 2018, Lima et al. 2019a). Of the total number (620) of the studies evaluated, most were of local interest, and the study period was up to two years' duration. In addition, there is an increasing trend to conduct research on new artificial reef materials, fishing productivity and coastal mitigation (Chen et al. 2015, López et al. 2016).

Throughout the five decades of artificial reef science evaluated here (1962–2018), ecological studies primarily investigated the capacity of the reefs for attraction of populations or biological assemblages associated with these reefs. Studies on animal behavior reported patterns of clustering, movement, and foraging of different commercially important species (Taylor et al. 2018) and focused on artificial reefs serving as shelters from predators and as breeding sites (Kimura and Munehara 2010, Orchard et al. 2018).

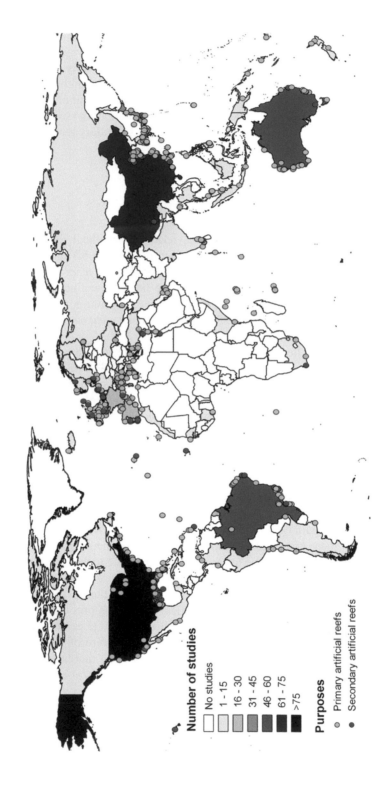

FIGURE 3.1 Distribution of studies on artificial reefs by country and general intent (i.e., primary and secondary artificial reefs).

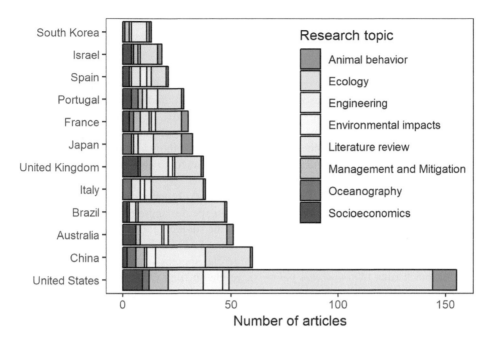

FIGURE 3.2 Number of articles per topic investigated according to major countries conducting artificial reef research.

Engineering studies addressed topics on the development of new artificial reefs materials such as biogenic matter and industrial waste (Chen et al. 2015, Huang et al. 2016, Mohamad et al. 2016). Other engineering studies addressed issues related to flow fields surrounding the reef (Miao and Xie 2007, Kim et al. 2014), improved artificial design (Kim et al. 2016, López et al. 2016), and settlement of the reef on the sea floor (Yun and Kim 2018).

Oceanographic studies often focused on biogeochemical, geomorphological, hydrodynamic, and nutrient-cycling aspects associated with artificial reefs (Falcão et al. 2009, Tiron et al. 2015). Other oceanographic-oriented publications focused on mapping artificial reefs using acoustic sonar to estimate reef volume, while others characterized the biota associated with reef structures (Raineault et al. 2013).

Studies that include socio-economic aspects associated with artificial reefs addressed issues related to their use as potential fishing areas and in promoting underwater tourism (Edney and Spennemann 2015, Belhassen et al. 2017). Additional investigations addressed evaluating their economic valuation and the environmental goods and services the reefs provide (Chen et al. 2013, Westerberg et al. 2013).

Published reviews on management and mitigation discussed issues related to the management of artificial reefs (Ng et al. 2013, Techera and Chandler 2015), carrying capacity (Murillas-Maza et al. 2013, Bush and Mol 2015), and the use of these structures in the restoration of aquatic environments (Guan et al. 2016). Research focused on environmental impacts addressed the negative effects of the deployment of artificial reefs (Aguilera et al. 2016).

OBSTACLES TO THE USE OF ARTIFICIAL REEFS

Artificial reef science has made significant advances in recent decades, especially regarding the attraction and/or production potential of these habitats (Smith et al. 2015, 2016, Glenn et al. 2017), the manufacture of structures with new materials (Chen et al. 2015, Dennis et al. 2018), and the development of different techniques for data analysis (Cresson et al. 2014, Becker et al. 2017, Reynolds et al. 2018). Conversely, studies on artificial reefs have encountered obstacles regarding: 1) impacts of non-inert materials in the environment; 2) gaps regarding socio-economic aspects; 3) the lack of integrated management studies; 4) long-term monitoring; and 5) uneven application of research (i.e., regional inequality of research efforts).

The use of non-inert materials, with the potential of high toxicity, exacerbates the negative impacts of primary reefs in aquatic environments (Figure 3.3). Tires as artificial reefs were considered advantageous in terms of resource recycling, low cost, and with high potential for species colonization (Russell 1975, Downing et al. 1985). More recently, the scientific community has considered the use of this material to be problematic, due to the physical instability of tires and the potential release (i.e., leaching) of chemical compounds into the water (such as petroleum derivatives, zinc, copper, formaldehyde, and acetone) that are harmful to marine biota (Collins et al. 2002, Sherman and Spieler 2006).

FADs are identified by the FAO (Food and Agricultural Organization of the United Nations) as an alternative to artisanal fishing and are the main multi-species aggregating structures in the Indian Ocean. However, the use of these structures has been the subject of discussion (Dagorn et al. 2013b, Bush and Mol 2015, Lopez et al. 2017). The expansion of FADs is due their ease of deployment, low cost, and their potential to attract important commercial species such as mahi-mahi (*Coryphaena hippurus*), skipjack tuna (*Katsuwonus pelamis*), yellowfin tuna (*Thunnus albacares*), and bigeye tuna (*Thunnus obesus*) (Dagorn et al. 2013a, Davies et al. 2014, Bell et al. 2018). However, several studies have noted the negative effects of FADs, as these structures may create opportunities for overfishing and facilitate the accidental capture of some vulnerable species (e.g., sharks, turtles, and cetaceans) (Leroy et al. 2013, Blasi et al. 2016, Tolentino-Zondervan et al. 2018).

Scuttled ships and shipwrecks have been used since World War II and provide habitats for attracting species and seascapes for recreational diving (Kirkbride-Smith et al. 2013, Krumholz and Brennan 2015). Similarly, offshore oil/gas platforms have been deactivated and reused as artificial reefs, as they often attract commercial fish species. Additionally, the high cost of their removal and transport of these structures (Love et al. 2012, Bond et al. 2018, Bull and Love 2019) often makes their deployment as artificial reefs an economically feasible alternative. Although several studies have determined that shipwrecks and offshore platforms increase seabed complexity and increase primary productivity (Muño-Pérez 2008, Love et al. 2012, Layman et al. 2016, Bond et al. 2018), other studies indicated that the excessive and unplanned use of these structures may result in risks for navigation, an increase in the mortality of some species, and an increase in the release of toxic materials into the oceans (Cripps and Aabel 2002, Garrido et al. 2015, Hamdan et al. 2018).

In addition to these structures, other secondary artificial reefs (e.g., port structures and wind energy foundations) have also been identified as facilitators of pathogen dissemination (Kendall et al. 2007, Villareal et al. 2007) as well as providing opportunities for the introduction and/or expansion of invasive species (Macneil and Platvoet 2013, Dong et al. 2018). Moreover, these structures may enable the accumulation of marine litter (Aguilera et al. 2016) and affect the behavior of seabirds (Petersen and Malm 2006, Punt et al. 2009). Despite criticisms regarding these structures, other studies have indicated the positive effects of these reefs on erosion control (Almeida 2017), seabed restoration (Langhamer et al. 2018), and shelter for several species (Harriague et al. 2013, Oricchio et al. 2016).

Although most studies have shown that operational aspects and materials engineering are as important for advancing the use of artificial reefs (Dafforn et al. 2015, Chapman et al. 2018), socio-economic aspects have been neglected by the scientific community (Lima et al. 2018). Studies on human populations affected by artificial reefs are still scarce, despite their economic importance to local communities (Evans et al. 2017, Macusi et al. 2017). The lack of dialog between decision-makers, the scientific community, and resource managers has generated conflicts of interest and has led to projects that often detached from local needs (Fletcher et al. 2011, Islam et al. 2014).

The low number of socio-economic studies has been due to the difficulties encountered by some researchers regarding developing studies with this approach, and resistance to the use of

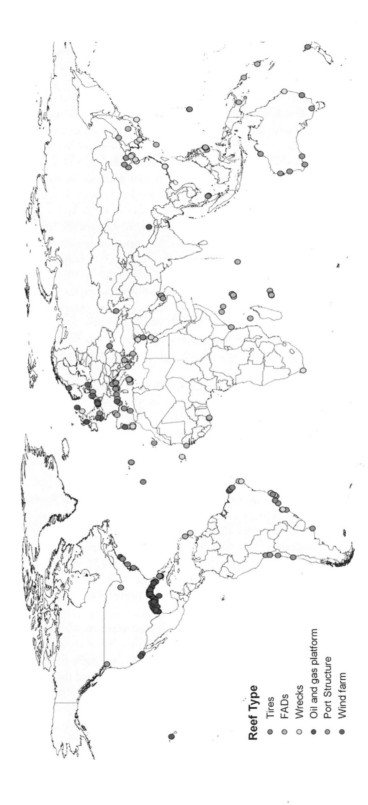

FIGURE 3.3 Global distribution of publications on artificial reefs considered here by generalized structures.

qualitative–quantitative ethnographic methods (Lima et al. 2019a). Opposition to the use of such methods has been based on issues such as an alleged lack of rigor in the collection of information, difficulty regarding defining the ideal number of informants capable of explaining a particular phenomenon, and questions regarding the validity of information based on the beliefs, values, and memories of the informants. Despite these presumed methodological limitations, the scope of human impacts on artificial reefs requires an integrated assessment that involves different social actors (e.g., fishers, divers) capable of providing information on reefs that are often overlooked by researchers and/or environmental agencies (Belhassen et al. 2017, Evans et al. 2017).

Artificial reef projects that disregard socio-economic effects tend to face greater difficulties in combining the demands of fishing resource exploitation and/or underwater tourism with the maintenance of artificial habitats (Andriesse 2018, Rouse et al. 2018, Lima et al. 2019b). In addition to these factors, the lack of clear study objectives and the use of methodologies that do not adequately address the hypotheses posed decrease the relevance of the final results because they are unable to adequately meet the needs of environmental managers (Feary et al. 2011).

Overall, artificial reef management has been characterized by specific initiatives for the installation of structures made of different materials with either open access to fishing activities or tourism exploitation (McLean et al. 2015, Firth et al. 2016). Although management studies have made significant progress over the past few years (Becker et al. 2018), the practical application of most of them has been limited, mainly because there is no methodological protocol that evaluates the efficiency of these structures from an ecological, social, and economic perspective and because of the difficulty of integrating interdisciplinary data. The absence of a unified method that fully analyzes the broad range of results from artificial reef studies has made it difficult to define the reference conditions of these habitats and to formulate global models of development and management in locations with artificial structures.

Initiatives such as the National Oceanic Policy of the USA, the Ocean Policy in Australia, and the European Union Water Framework Directive have emphasized the use of comprehensive management of aquatic environments, including guidelines on the strategic use of artificial reefs (Mead and Black 1999, Fabi et al. 2011). In countries without legislation regulating and providing guidelines on the deployment and monitoring of artificial reefs, there has been difficulty in maintaining the integrity of structures and efficiency of reefs in fishery production areas (Islam et al. 2014).

Although the guidelines for artificial reef management depend heavily on ecological and socio-economic concepts, the standards currently available for reef management are still subjective, complex, and difficult to access by managers. Another obstacle to management is related to the use of artificial reefs for protecting biodiversity (Pascaline et al. 2011). Some authors consider that areas with artificial structures can be used to form global networks of protected areas, functioning as buffer zones against anthropic impacts (Harmelin-Vivien et al. 2015), especially in areas with overfishing (Walker and Schlacher 2014). Globally, however, there have been few circumstances where areas with artificial reefs have been used as biodiversity protection zones (Francour et al. 2001, Pascaline et al. 2011).

Artificial reef studies have often been relegated to short-term monitoring, not exceeding two years. There has been a lack of constant and extended monitoring of these structures which limits the long-term quality of the information obtained. Moreover, the lack of long-term maintenance potentially leads to the burial of reef modules (Yun and Kim 2018) and the excessive accumulation of marine biofouling (Garrido et al. 2015). In addition, the short-term studies are often developed on a local scale and usually investigate the associated organisms (i.e., algae, epifauna, infauna, and fish) separately. Short-term, compartmentalized investigations, although essential for the analysis of faunal attraction and species colonization, are unable to detect slow and subtle long-term changes. This may lead to incorrect conclusions about the functioning of artificial reefs (Bortone et al. 2011, D'itri 2018). In contrast, long-term studies allow the detection of slow-acting processes and changes in species composition, revealing subtle and gradual biological trends (Santos et al. 2011, Kulaw et al. 2017).

Long-term and large-scale studies on artificial reefs, despite their infrequency at a global level, have been important for the development of this theme (Nicoletti et al. 2007, Becker et al. 2018). Studies based on global-scale, long-term series data provide more robust results, allowing a broader view that contributes to more effective management measures (Santos and Zalmon 2015, Kulaw et al. 2017). The lack of long-term studies at larger scales can be explained by the impediments to collaborations between different research institutions, the lack of sufficient funding, and the difficulty of coordinating simultaneous studies at different locations, as well as the fact that each country has its own regional problems and issues.

Despite the increased global interest in understanding how artificial reefs can function in resource management, most studies have been conducted in subtropical zones (Lee et al. 2018, Lima et al. 2019a). The regional inequality with regard to the number of studies is another obstacle to a broader assessment of reef complexes under different geographic, political, and socio-economic conditions (Ramos et al. 2011, 2011b, Sun et al. 2017). Considering that these reefs can help protect coastal areas, increase biodiversity, and serve to enhance fishing, monitoring them in different countries becomes important in any attempt to fill existing data gaps (Scarcella et al. 2015). In African, Latin American, and Southeast Asian countries, studies on artificial reefs are few, even though global biodiversity is concentrated in these regions. A lack of attention to investigations in high-diversity regions could result in serious environmental problems (Konan-Brou and Guiral 1994, Mbaru et al. 2018, Selfati et al. 2018). Global inequity in scientific development involves the historical, social, economic, and political aspects of each country. In addition, the disparity in the distribution of studies can be attributed to different emphasis of research in each country (Bulleri and Chapman 2010).

OVERCOMING CHALLENGES IN ARTIFICIAL REEF RESEARCH

In many ways, artificial reefs lend themselves, almost ideally, to studies of ecological processes, representing a powerful model for detecting temporal changes and the effects of environmental impacts. While some studies have focused on testing hypotheses about the effect of structures *in situ* (Kimura and Munehara 2010, Pascaline et al. 2011, Fariñas-Franco and Roberts 2014), much has also been learned through experimental studies conducted in controlled environments (Chen et al. 2015, Almeida 2017, Zheng et al. 2018). Artificial reefs have great potential to meet the many objectives for which they are intended; however, they still need to overcome the apparent disparity between theory and application. Although recent studies have contributed to the inclusion of new topics in the discussion about these reefs (Becker et al. 2018, Chapman et al. 2018, Bull and Love 2019), it is expected that future studies will incorporate interdisciplinary approaches that incorporate assessment methods that are consistent with meeting management objectives.

Incorporating alternative, low-impact materials in artificial reef construction is a trend of recent studies (Fletcher et al. 2011, Kirkbride-Smith et al. 2013, Bush and Mol 2015). Some studies on innovative materials have been conducted (Figure 3.4), mainly using biogenic materials (e.g., shell reefs) (Brown et al. 2014, Lv et al. 2016, Liu et al. 2017), industrial waste (e.g., green concrete—concrete that uses less energy and produces less carbon dioxide than normal concrete) (Bedoya-Gutierréz et al. 2016, Huang et al. 2016, Kim et al. 2016), and materials that are easy to form into a variety of shapes (e.g., geotextiles and fiberglass) (Loh et al. 2006, Fletcher et al. 2011, Kheawwongjan and Kim 2012).

Artificial reefs composed of rock (Davis and Smith 2017), clay (Brotto and Araujo 2001), bamboo (Konan-Brou and Guiral 1994), and timber (Yamamoto et al. 2014) have been used for to attract species to reefs because they are inexpensive, often occur in the natural environment, and do not release toxic substances. However, an increasing number of studies on reefs composed of bivalve molluscs (Brown et al. 2014, Walles et al. 2016) indicates an increasing trend in using this biogenic material (Figure 3.4) for environmental enrichment (i.e., increased substrate and increased attraction of biological resources). The impetus for using this material for reefs is based on recent challenges related to disposing of organic waste from the aquaculture industry. The industry, thus,

FIGURE 3.4 Global distribution of artificial reefs examined here according to new materials of composition.

provides a use for this waste that allows for increasing fishery resources with natural materials with a low environmental impact (Fariñas-Franco and Roberts 2014).

Material engineering for artificial reefs has made important advances in the formulation of new materials, such as the use of concrete mixed with various substances. Recently, industrial waste (Bedoya-Gutierréz et al. 2016, Liu et al. 2017) and calcium carbonate (Oyamada et al. 2008) is becoming incorporated into concrete structures, providing greater stability and chemical enrichment with low environmental impact. Fiberglass and geotextiles are becoming more frequently used in the construction of artificial reefs. Fiberglass is being used because it is chemically adequate and malleable enough to create modules with complex shapes (Loh et al. 2006, Foster et al. 2014). Geotextiles are being used because of their high capacity for erosion control and seabed modification (Mendonça et al. 2012).

Newer designs of artificial structures have improved their durability, strength, current flow, and biological attraction (Miao and Xie 2007, Kim et al. 2014). In addition, these new designs have minimized processes related to the submersion of the structures on the seabed (Yun and Kim 2018). An increase in studies evaluating the spatial configuration of the modules (Gatts et al. 2014) and the size of the reef complex (Layman et al. 2016) seems to be another trend in artificial reef science, because the results from these studies can help define sizes and arrangements of artificial habitats to increase biological resources with lower environmental impacts (Muño-Pérez 2008, López et al. 2016).

Advances in the use of new methods for data collection and analysis have occurred in artificial reef science; however, the expansion of the use of these new methods is often hindered by high operational costs. The use of remote underwater videos (e.g., Remotely Operated Vehicle— ROV and Baited Remote Underwater Video—BRUV) has produced information about the biological resources associated with reefs deployed at great depths or surveyed under adverse conditions (Ajemian et al. 2015, Becker et al. 2017). Sonar has been used to monitor artificial structures, facilitating the generation of information on the abundance, movement, and use of the reef by multiple species (Boswell et al. 2010).

Data obtained from sonars inserted in FADs has served to increase the attraction potential of these structures (Cillari et al. 2018, Groba et al. 2018). In the Pacific Islands, the use of FADs with sonar buoys has allowed access to oceanic fish resources by artisanal fishermen, thereby improving coastal food security (Bell et al. 2018). Stable isotopes and organic indicators have been widely used in studies on trophic ecology (Cheung et al. 2010, Cresson et al. 2014), molecular biology has been used to evaluate DNA changes in populations associated with artificial reefs (Salamone et al. 2016), and remote sensing by vessels has been used to evaluate, on a large scale, the selection of artificial reefs for use as fishing areas or underwater tourism (Wood et al. 2016).

The application of new analyses in computational engineering, ecological modeling, multivariate analyses, and meta-analysis has provided robust data on a large spatial-temporal scale and portends a trend in analytical methodology for the future (Campbell et al. 2011, Smith et al. 2017). However, there are still insufficient analyses on the resilience of artificial reefs to climate change.

Metrics used to evaluate the productivity of artificial reefs have been frequently based on the abundance, richness, size, biomass, and diversity of biological resources (Selfati et al. 2018, Cresson et al. 2019). However, considering the biological aspects while disregarding the socio-economic aspects can lead to biased conclusions about the function of these habitats (Carral et al. 2018, Lima et al. 2018). Studies conducted in Portugal (Ramos et al. 2011), the United Kingdom (Hooper et al. 2017), the Philippines (Macusi et al. 2017), Australia (Schaffer and Lawley 2012), and off the African coast (Moreno et al. 2007) have attempted to fill these gaps and have shown important advances in concrete models for the installation, participation, and co-management of artificial reefs.

Despite important advances in ecological approaches and performance monitoring, the lack of a standardized and global methodological protocol to assess the efficiency of these habitats in different regions has been an obstacle that needs to be overcome in future management studies (Becker et al. 2018, Bull and Love 2019). In response to this gap, multi-metric indices have emerged as a viable alternative because of the possibility of combining different information into a single or a few numerical values and because they are easy to sample, widely applicable and repeatable, consistent in terms of management, and easy to understand by the public (Borja 2005). Criticisms of the use of multi-metric indices usually revolve around the difficulty for one single value to represent the high complexity of ecological systems, in addition to the challenges for researchers in selecting metrics capable of detecting spatio-temporal variability in different environments (Borja and Dauer 2008).

In the future, interdisciplinary studies and long-term monitoring should be prioritized, especially in the evaluation of fishing potential and conservation in neotropical environments. The evaluation of multi-metric indices, based on global data (Career 2019, FAO 2019, FEOW 2019, Froese and Pauly 2019, GBIF 2019, IUCN 2019), and the use of long time-series has the potential for obtaining robust and reliable data for the management of artificial reefs. Studies conducted at broader geographic scales and longer time periods have shown the ecological role of artificial reefs and the potential of these structures in increasing fishery production (Leitão 2013). The benefits of such approaches are evident in the improvement of management projects and the functional analysis of artificial reefs in different ecological, economic, and social contexts (Fletcher et al. 2011, Ng et al. 2013). In addition, they also help support the co-participation of researchers, public agencies, and decision-makers in determining whether artificial reefs are an appropriate solution to local demands (Murillas-Maza et al. 2013).

Substantial changes in the current management of artificial reefs have been suggested in terms of implementation, monitoring, and comprehensive management of these areas (Kim and Kim 2008, Brochier et al. 2015). Future challenges imposed on the use and monitoring of artificial reefs in high seas include improvements in the decommissioning of oil/gas platforms and advances in the submersion of large structures (Techera and Chandler 2015, Hamdan et al. 2018, Bull and Love 2019). Studies in the Gulf of Mexico, off the California coast, and in the North Sea have been interesting models for the use and monitoring of these structures, which can be improved and implemented in other countries (Glenn et al. 2017, Rouse et al. 2018). As the management of oil/gas platforms and shipwrecks is often accorded to multiple legal jurisdictions and depends on the permission requirements of each country, researchers are encouraged to cooperate with government agencies to explore additional opportunities for these structures regarding the evaluation of the status of fish stocks, clarification of the uncertainty regarding "attraction vs production" and mitigation of biogeochemical impacts (Ajemian et al. 2015, Techera and Chandler 2015, Kirkbride-Smith et al. 2016, Kulaw et al. 2017).

The use of artificial reefs as part of marine protected areas has also been adopted by several countries (Francour et al. 2001, Muño-Pérez 2008). Strategies for conservation of the marine protected areas should include the full protection of artificial structures, preventing illegal catches by artisanal and industrial fisheries (Ramos et al. 2011, Tynyakov et al. 2017). Studies that map the occurrence and suppression of species, monitor fishing resources, and analyze the efficiency of reef complexes with and without protection should be encouraged (Seaman Jr. 2000, Andriesse 2018). A conservation program in these areas is just one of the initiatives that provides an option for increasing fishing resources (Bucaram et al. 2018). Conservation programs combined with large-scale monitoring and a reduction in fishing fleets also have the potential to halt the current decline of fisheries and provide a sustainable increase in fishing production (Smith et al. 2016).

Our analysis also indicates that the successful implementation of artificial reefs has been the result of the quality of prior planning and the continuous management of areas with artificial structures (Fabi et al. 2011, Kheawwongjan and Kim 2012). Plans should clearly specify objectives and technical assessments features. These should include the reason for implementation, the location and materials, the particularities of each ecosystem, and the total cost of implementation and maintenance (Kim and Kim 2008, Fletcher et al. 2011). Although there has been increasing recognition of the importance of integrated management, the lack of participation of all levels of society, the limited dissemination of information regarding artificial reefs through main communication media, and the lack of openness and connectivity of the research data comprise other obstacles that needs to be overcome (Vicente-Saez and Martinez-Fuentes 2018). Democratizing scientific knowledge and increasing public awareness regarding the potential of artificial reefs should be seen by researchers as a fundamental step in the development of research projects (Guenther and Joubert 2017). Scientific dissemination and information connectivity, in addition to increasing the impact of scientific research, is an important tool for disseminating knowledge about the functioning of artificial reefs and for promoting environmental goods and services provided by artificial reefs (Cooke et al. 2017).

In general, this review reveals some important trends in artificial reef science, such as a better understanding of settlement mechanisms, fishery production, and evaluations of environmental impacts from human activities. It is believed that progress depends on the development of low-impact materials, the use of new methods of analysis, an increase in the number of long-term studies at a broader geographical scale, the development of an evaluation protocol that includes socio-economic aspects, the development of conservation policies, and investment in scientific dissemination. Finally, it is hoped that this review will contribute not only to providing information on the obstacles in artificial reef science, but also to listing actions that can be taken to fill potential gaps in the development of future studies.

REFERENCES

Aguilera, M.A., R.R. Broitman, and M. Thiel. 2016. Artificial breakwaters as garbage bins: Structural complexity enhances anthropogenic litter accumulation in marine intertidal habitats. *Environ. Pollut.* 214:737–747. doi: 10.1016/j.envpol.2016.04.058

Ajemian, M.J., J.J. Wetz, B. Shipley-Lozano, J.D. Shively, and G.W. Stunz. 2015. An analysis of artificial reef fish community structure along the northwestern Gulf of Mexico shelf: Potential impacts of "rigs-to-reefs" programs. *PLOS ONE* 1:1–22. doi: 10.1371/journal.pone.0126354

Almeida, J.P.P.G.L. 2017. REEFS: An artificial reef for wave energy harnessing and shore protection – A new concept towards multipurpose sustainable solutions. *Renew. Energy* 114:817–829. doi: 10.1016/j.renene.2017.07.076

Andriesse, E. 2018. Persistent fishing amidst depletion, environmental and socio-economic vulnerability in Iloilo Province, the Philippines. *Ocean. Coast. Manag.* 157:130–137. doi: 10.1016/j.ocecoaman.2018.02.004

Becker, A., M.D. Taylor, H. Folpp, and M.B. Lowry. 2018. Managing the development of artificial reef systems: The need for quantitative goals. *Fish Fish.* 19(4):740–752. doi: 10.1111/faf.12288

Becker, A., M.D. Taylor, and M.B. Lowry. 2017. Monitoring of reef associated and pelagic fish communities on Australia's first purpose built offshore artificial reef. *ICES J. Mar. Sci.* 74(1):277–285. doi: 10.1093/icesjms/fsw133

Bedoya-Gutierréz, M.A., J.I. Tobón, T. Correa-Herrera, and J.D. Correa-Rendón. 2016. Evaluación biológica y fisicoquímica de un mortero como sustrato para la fabricación de arrecifes artificiales. *Bol. Cienc. Tierra* 40:55–63. doi: 10.15446/rbct.n40.55818

Belhassen, Y., M. Rousseau, J. Tynyakov, and N. Shashar. 2017. Evaluating the attractiveness and effectiveness of artificial coral reefs as a recreational ecosystem service. *J. Environ. Manag.* 203(1):448–456. doi: 10.1016/j.jenvman.2017.08.020

Bell, J.D., J. Albert, G. Amos, C. Arthur, M. Blanc, D. Bromhead, S.F. Heron, A.J. Hobday, A. Hunt, D. Itano, P.A.S. James, P. Lehodey, G. Liu, S. Nicol, J. Potemra, G. Reygondeau, J. Rubani, J. Scutt Phillips, I. Senina, and W. Sokimi. 2018. Operationalising access to oceanic fisheries resources by small-scale fishers to improve food security in the Pacific Islands. *Mar. Policy* 88:315–322. doi: 10.1016/j.marpol.2017.11.008

Blasi, M.F., F. Roscioni, and D. Mattei. 2016. Interaction of loggerhead turtles (*Caretta caretta*) with traditional fish aggregating devices (FADs) in the Mediterranean Sea. *Herpetol. Conserv. Biol.* 11:386–401.

Bohnsack, J.A. 1989. Are high densities of fishes at artificial reefs the result of habitat limitation or behavioral preference? *Bull. Mar. Sci.* 44:631–645.

Bond, T., J.C. Partridge, M.D. Taylor, T.F. Cooper, and D.L. Mclean. 2018. The influence of depth and a subsea pipeline on fish assemblages and commercially fished species. *PLOS ONE* 13(11):1–33. doi: 10.1371/journal.pone.0207703

Borja, Á. 2005. The European water framework directive: A challenge for nearshore, coastal and continental shelf research. *Cont. Shelf Res.* 25(14):1768–1783. doi: 10.1016/j.csr.2005.05.004

Borja, A., and D.M. Dauer. 2008. Assessing the environmental quality status in estuarine and coastal systems: Comparing methodologies and indices. *Ecol. Indic.* 8(4):331–337. doi: 10.1016/j.ecolind.2007.05.004

Bortone, S.A., F.P. Brandini, G. Fabi, and S. Otake. 2011. *Artificial Reefs in Fisheries Management*. Boca Raton: CRC Press/Taylor & Francis.

Boswell, K.M., R.J. David Wells, J.H. Cowan, and C.A. Wilson. 2010. Biomass, density, and size distributions of fishes associated with a large-scale artificial reef complex in the Gulf of Mexico. *Bull. Mar. Sci.* 86(4):879–889. doi: 10.5343/bms.2010.1026

Brickhill, M.J., S.Y. Lee, and R.M. Connolly. 2005. Fishes associated with artificial reefs: Attributing changes to attraction or production using novel approaches. *J. Fish Biol.* 67:53–71. doi: 10.1111/j.0022-1112.2005.00915.x

Brochier, T., P. Auger, N. Thiam, M. Sow, S. Diouf, H. Sloterdijk, and P. Brehmer. 2015. Implementation of artificial habitats: Inside or outside the marine protected areas? Insights from a mathematical approach. *Ecol. Modell.* 297:98–106. doi: 10.1016/j.ecolmodel.2014.10.034

Brotto, D.S., and F.G. Araujo. 2001. Habitat selection by fish in an artificial reef in Ilha Grande Bay, Brazil. *Braz. Arch. Biol. Technol.* 44(3):319–324. doi: 10.1590/S1516-89132001000300015

Brown, L.A., J.N. Furlong, K.M. Brown, and M.K. La Peyre. 2014. Oyster reef restoration in the northern Gulf of Mexico: Effect of artificial substrate and age on nekton and benthic macroinvertebrate assemblage use. *Restor. Ecol.* 22(2):214–222. doi: 10.1111/rec.12071

Bucaram, S.J., A. Hearn, A.M. Trujillo, W. Rentería, R.H. Bustamante, G. Morán, G. Reck, and J.L. García. 2018. Assessing fishing effects inside and outside an MPA: The impact of the Galapagos Marine Reserve on the Industrial pelagic tuna fisheries during the first decade of operation. *Mar. Policy* 87:212–225. doi: 10.1016/j.marpol.2017.10.002

Bull, A.S., and M.S. Love. 2019. Worldwide oil and gas platform decommissioning: A review of practices and reefing options. *Ocean. Coast. Manag.* 168:274–306. doi: 10.1016/j.ocecoaman.2018.10.024

Bulleri, F., and M.G. Chapman. 2010. The introduction of coastal infrastructure as a driver of change in marine environments. *J. Appl. Ecol.* 47(1):26–35. doi: 10.1111/j.1365-2664.2009.01751.x

Bush, S.R., and A.P.J. Mol. 2015. Governing in a placeless environment: Sustainability and fish aggregating devices. *Environ. Sci. Policy* 53:27–37. doi: 10.1016/j.envsci.2014.07.016

Campbell, M.D., K. Rose, K. Boswell, and J. Cowan. 2011. Individual-based modeling of an artificial reef fish community: Effects of habitat quantity and degree of refuge. *Ecol. Modell.* 222(23–24):3895–3909. doi: 10.1016/j.ecolmodel.2011.10.009

Career, E. 2019. World register of marine species. http://www.marinespecies.org/index.php (accessed October 19, 2019).

Carral, L., J.C. Alvarez-Feal, J. Tarrio-Saavedra, M.J. Rodriguez Guerreiro, and J.A. Fraguela. 2018. Social interest in developing a green modular artificial reef structure in concrete for the ecosystems of the Galician Rías. *J. Clean. Prod.* 172:1881–1898. doi: 10.1016/j.jclepro.2017.11.252

Chapman, M.G., A.J. Underwood, and M.A. Browne. 2018. An assessment of the current usage of ecological engineering and reconciliation ecology in managing alterations to habitats in urban estuaries. *Ecol. Eng.* 120:560–573. doi: 10.1016/j.ecoleng.2017.06.050

Chen, C., T. Ji, Y. Zhuang, and X. Lin. 2015. Workability, mechanical properties and affinity of artificial reef concrete. *Constr. Build. Mater.* 98:227–236. doi: 10.1016/j.conbuildmat.2015.05.109

Chen, J.L., C.T. Chuang, R.Q. Jan, L.C. Liu, and M.S. Jan. 2013. Recreational benefits of ecosystem services on and around artificial reefs: A case study in Penghu, Taiwan. *Ocean. Coast. Manag.* 85:58–64. doi: 10.1016/j.ocecoaman.2013.09.005

Cheung, S.G., H.Y. Wai, and P.K.S. Shin. 2010. Fatty acid profiles of benthic environment associated with artificial reefs in subtropical Hong Kong. *Mar. Pollut. Bull.* 60(2):303–308. doi: 10.1016/j.marpolbul.2009.12.001

Cillari, T., A. Allegra, F. Andaloro, M. Gristina, G. Milisenda, and M. Sinopoli. 2018. The use of echo-sounder buoys in Mediterranean Sea: A new technological approach for a sustainable FADs fishery. *Ocean. Coast. Manag.* 152:70–76. doi: 10.1016/j.ocecoaman.2017.11.018

Collins, K.J., A.C. Jensen, J.J. Mallinson, V. Roenelle, and I.P. Smith. 2002. Environmental impact assessment of a scrap tyre artificial reef. *ICES J. Mar. Sci.* 59:243–249. doi: 10.1006/jmsc.2002.1297

Cooke, S.J., A.J. Gallagher, N.M. Sopinka, V.M. Nguyen, R.A. Skubel, N. Hammerschlag, S. Boon, N. Young, and A.J. Danylchuk. 2017. Considerations for effective science communication. *Facets* 2(1):233–248. doi: 10.1139/facets-2016-0055

Cresson, P., L. Le Direach, E. Rouanet, E. Goberville, P. Astruch, M. Ourgaud, and M. Harmelin-Vivien. 2019. Functional traits unravel temporal changes in fish biomass production on artificial reefs. *Mar. Environ. Res.* 145:137–146. doi: 10.1016/j.marenvres.2019.02.018

Cresson, P., S. Ruitton, M. Ourgaud, and M. Harmelin-Vivien. 2014. Contrasting perception of fish trophic level from stomach content and stable isotope analyses: A Mediterranean artificial reef experience. *J. Exp. Mar. Bio. Ecol.* 452:54–62. doi: 10.1016/j.jembe.2013.11.014

Cripps, S.J., and J.P. Aabel. 2002. Environmental and socio-economic impact assessment of Ekoreef, a multiple platform rigs-to-reefs development. *ICES J. Mar. Sci.* 59:300–308. doi: 10.1006/jmsc.2002.1293

D'itri, F.M. 2018. *Artificial Reefs - Marine and Freshwater Applications.* Boca Raton: CRC Press/Taylor & Francis.

Dafforn, K.A., T.M. Glasby, L. Airoldi, N.K. Rivero, M. Mayer-Pinto, and E.L. Johnston. 2015. Marine urbanization: An ecological framework for designing multifunctional artificial structures. *Front. Ecol. Environ.* 13(2):82–90. doi: 10.1890/140050

Dagorn, L., N. Bez, T. Fauvel, and E. Walker. 2013a. How much do fish aggregating devices (FADs) modify the floating object environment in the ocean? *Fish. Oceanogr.* 22(3):147–153. doi: 10.1111/fog.12014

Dagorn, L., K.N. Holland, V. Restrepo, and G. Moreno. 2013b. Is it good or bad to fish with FADs? What are the real impacts of the use of drifting FADs on pelagic marine ecosystems? *Fish Fish.* 14(3):391–415. doi: 10.1111/j.1467-2979.2012.00478.x

Davies, T.K., C.C. Mees, and E.J. Milner-Gulland. 2014. The past, present and future use of drifting fish aggregating devices (FADs) in the Indian Ocean. *Mar. Policy* 45:163–170. doi: 10.1016/j.marpol.2013.12.014

Davis, T.R., and S.D.A. Smith. 2017. Proximity effects of natural and artificial reef walls on fish assemblages. *Reg. Stud. Mar. Sci.* 9:17–23. doi: 10.1016/j.rsma.2016.10.007

Dennis, H.D., A.J. Evans, A.J. Banner, and P.J. Moore. 2018. Reefcrete: Reducing the environmental footprint of concretes for eco-engineering marine structures. *Ecol. Eng.* 120:668–678. doi: 10.1016/j.ecoleng.2017.05.031

Dong, Z., L. Wang, T. Sun, Q. Liu, and Y. Sun. 2018. Artificial reefs for sea cucumber aquaculture confirmed as settlement substrates of the moon jellyfish *Aurelia coerulea. Hydrobiologia* 818(1):223–234. doi: 10.1007/s10750-018-3615-y

Downing, N., R.A. Tubb, C.R. El-Zahr, and R.E. McClure. 1985. Artificial reefs in Kuwait, Northern Arabian Gulf. *Bull. Mar. Sci.* 37:157–178.

Edney, J., and D.H.R. Spennemann. 2015. Can artificial reef wrecks reduce diver impacts on shipwrecks? the management dimension. *J. Marit. Archaeol.* 10(2):141–157. doi: 10.1007/s11457-015-9140-5

Evans, A.J., B. Garrod, L.B. Firth, S.J. Hawkins, E.S. Morris-Webb, H. Goudge, and P.J. Moore. 2017. Stakeholder priorities for multi-functional coastal defence developments and steps to effective implementation. *Mar. Policy* 75:143–155. doi: 10.1016/j.marpol.2016.10.006

Fabi, G., A. Spagnolo, D. Bellan-Santini, E. Charbonnel, B.A. Çiçek, J.J.G. García, A.C. Jensen, A. Kallianiotis, and M.N. dos Santos. 2011. Overview on artificial reefs in Europe. *Braz. J. Oceanogr.* 59(spe1):155–166. doi: 10.1590/S1679-87592011000500017

FAO. 2019. Food and Agriculture Organization of the United Nations - Fisheries and Aquaculture Department. http://www.fao.org/fishery/statistics/en (accessed October 19, 2019)

Falcão, M., M.N. Santos, T. Drago, D. Serpa and C. Monteiro. 2009. Effect of artificial reefs (southern Portugal) on sediment-water transport of nutrients: Importance of the hydrodynamic regime. *Estuar. Coast. Shelf. S.* 83:451–459. doi: 10.1016/j.ecss.2009.04.028

Fariñas-Franco, J.M., and D. Roberts. 2014. Early faunal successional patterns in artificial reefs used for restoration of impacted biogenic habitats. *Hydrobiologia* 727(1):75–94. doi: 10.1007/s10750-013-1788-y

Feary, D.A., J.A. Burt, and A. Bartholomew. 2011. Artificial marine habitats in the Arabian Gulf: Review of current use, benefits and management implications. *Ocean. Coast. Manag.* 54(10):742–749. doi: 10.1016/j.ocecoaman.2011.07.008

FEOW. 2019. Freshwater ecoregions of the world. http://www.feow.org (accessed October 19, 2019).

Firth, L.B., A.M. Knights, D. Bridger, A.J. Evans, N. Mieszkowska, P.J. Moore, N.E. O'Connor, E.V. Sheehan, R.C. Thompson, and S.J. Hawkins. 2016. Ocean sprawl: Challenges and opportunities for biodiversity management in a changing world. *Oceanogr. Mar. Biol. Annu. Rev.* 54:193–269. doi: 10.1201/9781315368597

Fletcher, S., P. Bateman, and A. Emery. 2011. The governance of the Boscombe Artificial Surf Reef, UK. *Land Use Policy* 28(2):395–401. doi: 10.1016/j.landusepol.2010.08.002

Foster, S., D.A. Smale, J. How, S. de Lestang, A. Brearley, and G.A. Kendrick. 2014. Regional-scale patterns of mobile invertebrate assemblage structure on artificial habitats off Western Australia. *J. Exp. Mar. Bio. Ecol.* 453:43–53. doi: 10.1016/j.jembe.2013.12.015

Francour, P., J.-G. Harmelin, D. Pollard, and S. Sartoretto. 2001. A review of marine protected areas in the northwestern Mediterranean region: Siting, usage, zonation and management. *Aquat. Conserv. Mar. Freshw. Ecosystems* 11:155–188. doi: 10.1002/aqc.442

Froese, R., and D. Pauly. 2019. FishBase. World Wide Web electronic publication. www.fishbase.org (accessed October 24, 2019).

Garrido, P.H.L., J. González-Sánchez, and E.E. Briones. 2015. Fouling communities and degradation of archeological metals in the coastal sea of the Southwestern Gulf of Mexico. *Biofouling* 31(5):405–416. doi: 10.1080/08927014.2015.1048433

Gatts, P.V., M.A.L. Franco, L.N. Santos, D.F. Rocha, and I.R. Zalmon. 2014. Influence of the artificial reef size configuration on transient ichthyofauna - Southeastern Brazil. *Ocean. Coast. Manag.* 98:111–119. doi: 10.1016/j.ocecoaman.2014.06.022

GBIF. 2019. Global Biodiversity Information Facility. http://www.gbif.org (accessed October 24, 2019).

Glenn, H.D., J.H. Cowan, J.E. Powers. 2017. A comparison of red snapper reproductive potential in the northwestern Gulf of Mexico: Natural versus artificial habitats. *Mar. Coast. Fish.* 9(1):139–148. doi: 10.1080/19425120.2017.1282896

Groba, C., A. Sartal, and X.H. Vázquez. 2018. Integrating forecasting in metaheuristic methods to solve dynamic routing problems: Evidence from the logistic processes of tuna vessels. *Eng. Appl. Artif. Intell.* 76:55–66. doi: 10.1016/j.engappai.2018.08.015

Guan, M.L., T. Zheng, and X.Y. You. 2016. Ecological rehabilitation prediction of enhanced key-food-web offshore restoration technique by wall roughening. *Ocean. Coast. Manag.* 128:1–9. doi: 10.1016/j.ocecoaman.2016.04.008

Guenther, L., and M. Joubert. 2017. Science communication as a field of research: Identifying trends, challenges and gaps by analysing research papers. *Sci. Commun.* 16(2):1–19. doi: 10.22323/2.16020202

Hamdan, L.J., J.L. Salerno, A. Reed, S.B. Joye, and M. Damour. 2018. The impact of the deepwater horizon blowout on historic shipwreck-associated sediment microbiomes in the northern Gulf of Mexico. *Sci. Rep.* 8(1):1–14. doi: 10.1038/s41598-018-27350-z

Harmelin-Vivien, M., J.M. Cottalorda, J.M. Dominici, J.-G. Harmelin, L. Le Diréach, and S. Ruitton. 2015. Effects of reserve protection level on the vulnerable fish species *Sciaena umbra* and implications for fishing management and policy. *Glob. Ecol. Conserv.* 3:279–287. doi: 10.1016/j.gecco.2014.12.005

Harriague, A.C., C. Misic, I. Valentini, E. Polidori, G. Albertelli, and A. Pusceddu. 2013. Meio- and macrofauna communities in three sandy beaches of the northern Adriatic Sea protected by artificial reefs. *Chem. Ecol.* 29:181–195. doi: 10.1080/02757540.2012.704911

Hooper, T., C. Hattam, and M. Austen. 2017. Recreational use of offshore wind farms: Experiences and opinions of sea anglers in the UK. *Mar. Policy* 78:55–60. doi: 10.1016/j.marpol.2017.01.013

Huang, X., Z. Wang, Y. Liu, W. Hu, and W. Ni. 2016. On the use of blast furnace slag and steel slag in the preparation of green artificial reef concrete. *Constr. Build. Mater.* 112:241–246. doi: 10.1016/j.conbuildmat.2016.02.088

Islam, G.M.N., K.M. Noh, S.F. Sidique, and A.F.M. Noh. 2014. Economic impact of artificial reefs: A case study of small scale fishers in Terengganu, Peninsular Malaysia. *Fish. Res.* 151:122–129. doi: 10.1016/j.fishres.2013.10.018

IUCN. 2019. The IUCN red list of threatened species. http://www.iucnredlist.org (accessed November 5, 2019).

Kendall, L.R., J.W. Ewart, P.N. Ulrich, and A.G. Marsh. 2007. Low incidence and limited effect of the oyster pathogen dermo (*Perkinsus marinus*) on an artificial reef in Delaware's Inland Bays. *Estuaries Coasts* 30(1):154–162. doi: 10.1007/BF02782975

Kheawwongjan, A., and D.S. Kim. 2012. Present status and prospects of artificial reefs in Thailand. *Ocean. Coast. Manag.* 57:21–33. doi: 10.1016/j.ocecoaman.2011.11.001

Kim, C.G., and H.S. Kim. 2008. Post-placement management of artificial reefs in Korea. *Fisheries* 33(2):61–68. doi: 10.1577/1548-8446-33.2.61

Kim, D., J. Woo, H.S. Yoon, and W.B. Na. 2014. Wake lengths and structural responses of Korean general artificial reefs. *Ocean Eng.* 92:83–91. doi: 10.1016/j.oceaneng.2014.09.040

Kim, D., J. Woo, H.S. Yoon, and W.B. Na. 2016. Efficiency, tranquility and stability indices to evaluate performance in the artificial reef wake region. *Ocean Eng.* 122:253–261. doi: 10.1016/j.oceaneng.2016.06.030

Kimura, M.R., and H. Munehara. 2010. The disruption of habitat isolation among three Hexagrammos species by artificial habitat alterations that create mosaic-habitat. *Ecol. Res.* 25(1):41–50. doi: 10.1007/s11284-009-0624-3

Kirkbride-Smith, A.E., P.M. Wheeler, and M.L. Johnson. 2013. The relationship between diver experience levels and perceptions of attractiveness of artificial reefs - Examination of a potential management tool. *PLOS ONE* 8(7):1–10. doi: 10.1371/journal.pone.0068899

Kirkbride-Smith, A.E., P.M. Wheeler, and M.L. Johnson. 2016. Artificial reefs and marine protected areas: A study in willingness to pay to access Folkestone Marine Reserve, Barbados, West Indies. *PeerJ* 4:1–32. doi: 10.7717/peerj.2175

Konan-Brou, A.A., and D. Guiral. 1994. Available algal biomass in tropical brackish water artificial habitats. *Aquaculture* 119(2–3):175–190. doi: 10.1016/0044-8486(94)90174-0

Krumholz, J.S., and M.L. Brennan. 2015. Fishing for common ground: Investigations of the impact of trawling on ancient shipwreck sites uncovers a potential for management synergy. *Mar. Policy* 61:127–133. doi: 10.1016/j.marpol.2015.07.009

Kulaw, D.H., J.H. Cowan, and M.W. Jackson. 2017. Temporal and spatial comparisons of the reproductive biology of northern Gulf of Mexico (USA) red snapper (*Lutjanus campechanus*) collected a decade apart. *PLOS ONE* 12(3):1–39. doi: 10.1371/journal.pone.0172360

Langhamer, O., T.G. Dahlgren, and G. Rosenqvist. 2018. Effect of an offshore wind farm on the viviparous eelpout: Biometrics, brood development and population studies in Lillgrund, Sweden. *Ecol. Indic.* 84:1–6. doi: 10.1016/j.ecolind.2017.08.035

Layman, C.A., J.E. Allgeier, and C.G. Montaña. 2016. Mechanistic evidence of enhanced production on artificial reefs: A case study in a Bahamian seagrass ecosystem. *Ecol. Eng.* 95:574–579. doi: 10.1016/j.ecoleng.2016.06.109

Lee, M.O., S. Otake, and J.K. Kim. 2018. Transition of artificial reefs (ARs) research and its prospects. *Ocean. Coast. Manag.* 154:55–65. doi: 10.1016/j.ocecoaman.2018.01.010

Leitão, F. 2013. Artificial reefs: From ecological processes to fishing enhancement tools. *Braz. J. Oceanogr.* 61(1):77–81. doi: 10.1590/S1679-87592013000100009

Leroy, B., J.S. Phillips, S. Nicol, G.M. Pilling, S. Harley, D. Bromhead, S. Hoyle, S. Caillot, V. Allain, and J. Hampton. 2013. A critique of the ecosystem impacts of drifting and anchored FADs use by purse-seine tuna fisheries in the Western and Central Pacific Ocean. *Aquat. Living Resour.* 26(1):49–61. doi: 10.1051/alr/2012033

Lima, J.S., I.R. Zalmon, and M. Love. 2019a. Overview and trends of ecological and socioeconomic research on artificial reefs. *Mar. Environ. Res.* 145:81–96. doi: 10.1016/j.marenvres.2019.01.010

Lima, J.S., C.A. Zappes, A.P.M. Di Beneditto, and I.R. Zalmon. 2019b. Ethnoecology and socioeconomic around an artificial reef: The case of artisanal fisheries from southeastern Brazil. *Biota Neotrop.* 19(2):1–13. doi: 10.1590/1676-0611-bn-2018-0620

Lima, J.S., C.A. Zappes, A.P.M. Di Beneditto, and I.R. Zalmon. 2018. Artisanal fisheries and artificial reefs on the southeast coast of Brazil: Contributions to research and management. *Ocean. Coast. Manag.* 163:372–382. doi: 10.1016/J.OCECOAMAN.2018.07.018

Lindberg, W.J. 1997. Can science resolve the attraction-production issue? *Fisheries.* 22:10–16. doi: 10.1577/1548-8446-22-4

Liu, G., W.T. Li, and X. Zhang. 2017. Assessment of the benthic macrofauna in an artificial shell reef zone in Shuangdao Bay, Yellow Sea. *Mar. Pollut. Bull.* 114(2):778–785. doi: 10.1016/j.marpolbul.2016.11.004

Loh, T.L., J.T.I. Tanzil, and L.M. Chou. 2006. Preliminary study of community development and scleractinian recruitment on fibreglass artificial reef units in the sedimented waters of Singapore. *Aquat. Conserv. Mar. Freshw. Ecosyst.* 16(1):61–76. doi: 10.1002/aqc.701

López, I., H. Tinoco, L. Aragonés, and J. García-Barba. 2016. The multifunctional artificial reef and its role in the defence of the Mediterranean coast. *Sci. Total Environ.* 550:910–923. doi: 10.1016/j.scitotenv.2016.01.180

Lopez, J., G. Moreno, L. Ibaibarriaga, and L. Dagorn. 2017. Diel behaviour of tuna and non-tuna species at drifting fish aggregating devices (DFADs) in the Western Indian Ocean, determined by fishers' echo-sounder buoys. *Mar. Biol.* 164:1–16. doi: 10.1007/s00227-017-3075-3

Love, M.S., M. Nishimoto, C. Clark, and D. Schroeder. 2012. Recruitment of young-of-the-year fishes to natural and artificial offshore structure within central and southern California waters, 2008–2010. *Bull. Mar. Sci.* 88:863–882. doi: 10.5343/bms.2011.1101

Lv, W., Y. Huang, Z. Liu, Y. Yang, B. Fan, and Y. Zhao. 2016. Application of macrobenthic diversity to estimate ecological health of artificial oyster reef in Yangtze Estuary, China. *Mar. Pollut. Bull.* 103(1–2):137–143. doi: 10.1016/j.marpolbul.2015.12.029

Macneil, C., and D. Platvoet. 2013. Could artificial structures such as fish passes facilitate the establishment and spread of the "killer shrimp" *Dikerogammarus villosus* (Crustacea: Amphipoda) in river systems? *Aquat. Conserv. Mar. Freshw. Ecosyst.* 23:667–677. doi: 10.1002/aqc.2337

Macusi, E.D., N.A.S. Abreo, and R.P. Babaran. 2017. Local ecological knowledge (LEK) on fish behavior around anchored FADs: The case of tuna purse seine and ringnet fishers from Southern Philippines. *Front. Mar. Sci.* 4:1–13. doi: 10.3389/fmars.2017.00188

Mbaru, E.K., D. Sigana, R.K. Ruwa, E.M. Mueni, C.K. Ndoro, E.N. Kimani, and B. Kaunda-Arara. 2018. Experimental evaluation of influence of FADs on community structure and fisheries in coastal Kenya. *Aquat. Living Resour.* 31(6):1–12. doi: 10.1051/alr/2017045

McLean, M., E.F. Roseman, J.J. Pritt, G. Kennedy, and B.A. Manny. 2015. Artificial reefs and reef restoration in the Laurentian Great Lakes. *J. Gr Lakes Res.* 41(1):1–8. doi: 10.1016/j.jglr.2014.11.021

Mead, S., and K. Black. 1999. A multipurpose, artificial reef at Mount Maunganui beach, New Zealand. *Coast. Manag.* 27:355–365. doi: 10.1080/089207599263767

Mendonça, A., C.J. Fortes, R. Capitão, M. da G. Neves, T. Moura, and J.S. Antunes do Carmo. 2012. Wave hydrodynamics around a multi-functional artificial reef at Leirosa. *J. Coast. Conserv.* 16(4):543–553. doi: 10.1007/s11852-012-0196-1

Miao, Z.Q., and Y.H. Xie. 2007. Effects of water-depth on hydrodynamic force of artificial reef. *J. Hydrodyn.* 19(3):372–377. doi: 10.1016/S1001-6058(07)60072-9

Mohamad, N., A.A.A. Samad, W.I. Goh, H. Monica, and F. Hasbullah. 2016. Nutrient leach from concrete artificial reef incorporating with organic material. *J. Teknol.* 78(5):23–27. doi: 10.11113/jt.v78.8231

Moher, D., A. Liberati, J. Tetzlaff, D.G. Altman, D. Altman, G. Antes, D. Atkins, V. Barbour, N. Barrowman, J.A. Berlin, J. Clark, M. Clarke, D. Cook, R. D'Amico, J.J. Deeks, P.J. Devereaux, K. Dickersin, M. Egger, E. Ernst, P.C. Gøtzsche, J. Grimshaw, G. Guyatt, J. Higgins, J.P.A. Ioannidis, J. Kleijnen, T. Lang, N. Magrini, D. McNamee, L. Moja, C. Mulrow, M. Napoli, A. Oxman, B. Pham, D. Rennie, M. Sampson, K.F. Schulz, P.G. Shekelle, D. Tovey, and P. Tugwell. 2009. Preferred reporting items for systematic reviews and meta-analyses: The PRISMA statement. *PLOS Med.* 6(7):1–6. doi: 10.1371/journal.pmed.1000097

Moreno, G., L. Dagorn, G. Sancho, D. García, and D. Itano. 2007. Using local ecological knowledge (LEK) to provide insight on the tuna purse seine fleets of the Indian Ocean useful for management. *Aquat. Living Resour.* 20(4):367–376. doi: 10.1051/alr:2008014

Muño-Pérez, J. 2008. Artificial reefs to improve and protect fishing grounds. *Recent Pat. Eng.* 2(2):80–86. doi: 10.2174/187221208784705260

Murillas-Maza, A., G. Moreno, and J. Murua. 2013. A socio-economic sustainability indicator for the Basque tropical tuna purse-seine fleet with a FAD fishing strategy. *Econ. Agrar. Recur. Nat.* 13(2):5–31. doi: 10.7201/earn.2013.02.01

Ng, K., M.R. Phillips, H. Calado, P. Borges, and F. Veloso-Gomes. 2013. Seeking harmony in coastal development for small islands: Exploring multifunctional artificial reefs for São Miguel Island, the Azores. *Appl. Geogr.* 44:99–111. doi: 10.1016/j.apgeog.2013.07.013

Nicoletti, L., S. Marzialetti, D. Paganelli, and G.D. Ardizzone. 2007. Long-term changes in a benthic assemblage associated with artificial reefs. *Hydrobiologia* 580(1):233–240. doi: 10.1007/s10750-006-0450-3

Orchard, D.S.E., M.J.H. Hickford, and D.R. Schiel. 2018. Use of artificial habitats to detect spawning sites for the conservation of *Galaxias maculatus*, a riparian-spawning fish. *Ecol. Indic.* 91:617–625. doi: 10.1016/j.ecolind.2018.03.061

Oricchio, F.T., G. Pastro, E.A. Vieira, A.A.V. Flores, F.Z. Gibran, and G.M. Dias. 2016. Distinct community dynamics at two artificial habitats in a recreational marina. *Mar. Environ. Res.* 122:85–92. doi: 10.1016/j.marenvres.2016.09.010

Oyamada, K., M. Tsukidate, K. Watanabe, T. Takahashi, T. Isoo, and T. Terawaki. 2008. A field test of porous carbonated blocks used as artificial reef in seaweed beds of *Ecklonia cava*. *J. Appl. Phycol.* 20(5):863–868. doi: 10.1007/s10811-008-9332-6

Pascaline, B., S. Catherine, E. Charbonnel, and P. Francour. 2011. Monitoring of the artificial reef fish assemblages of Golfe Juan marine protected area (France, North-Western Mediterranean). *Braz. J. Oceanogr.* 59(spe1):167–176. doi: 10.1590/S1679-87592011000500018

Petersen, J.K., and T. Malm. 2006. Offshore windmill farms: Threats to or possibilities for the marine environment. *Ambio A J. Hum. Environ.* 35(2):75–80. doi: 10.1579/0044-7447(2006)35[75:OWFTTO]2.0.CO;2

Punt, M.J., R.A. Groeneveld, E.C. van Ierland, and J.H. Stel. 2009. Spatial planning of offshore wind farms: A windfall to marine environmental protection? *Ecol. Econ.* 69(1):93–103. doi: 10.1016/j.ecolecon.2009.07.013

Raineault, N.A., A.C. Trembanis, D.C. Miller, and V. Capone. 2013. Interannual changes in seafloor surficial geology at an artificial reef site on the inner continental shelf. *Cont. Shelf Res.* 58:67–78. doi: 10.1016/j.csr.2013.03.008

Ramos, J., M.T. Oliveira, and M.N. Santos. 2011a. Stakeholder perceptions of decision-making process on marine biodiversity conservation on Sal Island (Cape Verde). *Braz. J. Oceanogr.* 59(spe1):95–105.

Ramos, J., M.N. Santos, D. Whitmarsh, and C.C. Monteiro. 2011b. Stakeholder analysis in the portuguese artificial reef context: winners and losers. *Braz. J. Oceanogr.* 59:133–143. doi: 10.1590/S1679-87592011000500015

Reynolds, E.M., J.H. Cowan, K.A. Lewis, and K.A. Simonsen. 2018. Method for estimating relative abundance and species composition around oil and gas platforms in the northern Gulf of Mexico, U.S.A. *Fish. Res.* 201:44–55. doi: 10.1016/j.fishres.2018.01.002

Rouse, S., A. Kafas, R. Catarino, and H. Peter. 2018. Commercial fisheries interactions with oil and gas pipelines in the North Sea: Considerations for decommissioning. *ICES J. Mar. Sci.* 75(1):279–286. doi: 10.1093/icesjms/fsx121

Russell, B.C. 1975. The development and dynamics of a small artificial reef community. *Helgol. Wiss.* 27(3):298–312.

Salamone, A.L., B.M. Robicheau, and A.K. Walker. 2016. Fungal diversity of marine biofilms on artificial reefs in the north-central Gulf of Mexico. *Bot. Mar.* 59(5):291–305. doi: 10.1515/bot-2016-0032

Santos, L.N., D.S. Brotto, and I.R. Zalmon. 2011. Assessing artificial reefs for fisheries management A 10-year assessment off northern coast of Rio de Janeiro. In: *Artificial Reefs in Fisheries Management*, ed. S.A. Bortone, F.P. Brandini, G. Fabi, S. Otake, 125–140. London/New York: CRC Press

Santos, L.N., and I.R. Zalmon. 2015. Long-term changes of fish assemblages associated with artificial reefs off the northern coast of Rio de Janeiro, Brazil. *J. Appl. Ichthyol.* 31:15–23. doi: 10.1111/jai.12947

Scarcella, G., F. Grati, L. Bolognini, F. Domenichetti, S. Malaspina, S. Manoukian, P. Polidori, A. Spagnolo, and G. Fabi. 2015. Time-series analyses of fish abundance from an artificial reef and a reference area in the central-Adriatic Sea. *J. Appl. Ichthyol.* 31:74–85. doi: 10.1111/jai.12952

Schaffer, V., and M. Lawley. 2012. An analysis of the networks evolving from an artificial reef development. *Curr. Issues Tour.* 15(5):497–503. doi: 10.1080/13683500.2011.638704

Seaman Jr., W. 2000. *Artificial Reef Evaluation with Application to Natural Marine Habitats.* Boca Raton: CRC Press/Taylor & Francis.

Selfati, M., N. El Oumari, P. Lenfant, A. Fontcuberta, G. Lecaillon, A. Mesfioui, P. Boissery, and H. Bazairi. 2018. Promoting restoration of fish communities using artificial habitats in coastal marinas. *Biol. Conserv.* 219:89–95. doi: 10.1016/j.biocon.2018.01.013

Sherman, R.L., and R.E. Spieler. 2006. Tires: Unstable materials for artificial reef construction. *WIT Trans. Ecol. Environ.* 88:215–223. doi: 10.2495/CENV060211

Smith, J.A., W.K. Cornwell, M.B. Lowry, and I.M. Suthers. 2017. Modelling the distribution of fish around an artificial reef. *Mar. Freshw. Res.* 68(10):1955–1964. doi: 10.1071/MF16019

Smith, J.A., M.B. Lowry, C. Champion, and I.M. Suthers. 2016. A designed artificial reef is among the most productive marine fish habitats: New metrics to address 'production versus attraction'. *Mar. Biol.* 163(9):1–8. doi: 10.1007/s00227-016-2967-y

Smith, J.A., M.B. Lowry, I.M. Suthers. 2015. Fish attraction to artificial reefs not always harmful: A simulation study. *Ecol. Evol.* 5(20):4590–4602. doi: 10.1002/ece3.1730

Streich, M.K., M.J. Ajemian, J.J. Wetz, J.D. Shively, J.B. Shipley, and G.W. Stunz. 2017. Effects of a new artificial reef complex on red snapper and the associated fish community: An evaluation using a before–after control–impact approach. *Mar. Coast. Fish.* 9(1):404–418. doi: 10.1080/19425120.2017.1347116

Sun, P., X. Liu, Y. Tang, W. Cheng, R. Sun, X. Wang, R. Wan, and M. Heino. 2017. The bio-economic effects of artificial reefs: Mixed evidence from Shandong, China. *ICES J. Mar. Sci.* 74(8):2239–2248. doi: 10.1093/icesjms/fsx058

Taylor, M.D., A. Becker, and M.B. Lowry. 2018. Investigating the functional role of an artificial reef within an Estuarine Seascape: a case study of yellowfin bream (*Acanthopagrus australis*). *Estuar. Coasts* 1:1–11. doi: 10.1007/s12237-018-0395-6

Techera, E.J., and J. Chandler. 2015. Offshore installations, decommissioning and artificial reefs: Do current legal frameworks best serve the marine environment? *Mar. Policy* 59:53–60. doi: 10.1016/j.marpol.2015.04.021

Tiron, R., F. Mallon, F. Dias, and E.G. Reynaud. 2015. The challenging life of wave energy devices at sea: A few points to consider. *Renew. Sust. Energ. Rev.* 43:1263–1272. doi: 10.1016/j.rser.2014.11.105

Tolentino-Zondervan, F., P. Berentsen, S.R. Bush, and Oude Lansink. 2018. FAD vs. free school: Effort allocation by Marine Stewardship Council compliant Filipino tuna purse seiners in the PNA. *Mar. Policy* 90:137–145. doi: 10.1016/j.marpol.2017.12.025

Tynyakov, J., M. Rousseau, M. Chen, O. Figus, Y. Belhassen, and N. Shashar. 2017. Artificial reefs as a means of spreading diving pressure in a coral reef environment. *Ocean. Coast. Manag.* 149:159–164. doi: 10.1016/j.ocecoaman.2017.10.008

Vicente-Saez, R., and C. Martinez-Fuentes. 2018. Open Science now : A systematic literature review for an integrated definition. *J. Bus. Res.* 88:428–436. doi: 10.1016/j.jbusres.2017.12.043

Villareal, T.A., S. Hanson, S. Qualia, E.L.E. Jester, H.R. Granade, and R.W. Dickey. 2007. Petroleum production platforms as sites for the expansion of ciguatera in the northwestern Gulf of Mexico. *Harmful Algae* 6(2):253–259. doi: 10.1016/j.hal.2006.08.008

Walker, S.J., and T.A. Schlacher. 2014. Limited habitat and conservation value of a young artificial reef. *Biodivers. Conserv.* 23(2):433–447. doi: 10.1007/s10531-013-0611-4

Walles, B., K. Troost, D. van den Ende, S. Nieuwhof, A.C. Smaal, and T. Ysebaert. 2016. From artificial structures to self-sustaining oyster reefs. *J. Sea Res.* 108:1–9. doi: 10.1016/j.seares.2015.11.007

Westerberg, V., J.B. Jacobsen, and R. Lifran. 2013. The case for offshore wind farms, artificial reefs and sustainable tourism in the French Mediterranean. *Tour. Manag.* 34:172–183. doi: 10.1016/j.tourman.2012.04.008

Wood, G., T.P. Lynch, C. Devine, K. Keller, and W. Figueira. 2016. High-resolution photo-mosaic time-series imagery for monitoring human use of an artificial reef. *Ecol. Evol.* 6(19):6963–6968. doi: 10.1002/ece3.2342

Yamamoto, K.C., C.E.de C. Freitas, J. Zuanon, and L.E. Hurd. 2014. Fish diversity and species composition in small-scale artificial reefs in Amazonian floodplain lakes: Refugia for rare species? *Ecol. Eng.* 67:165–170. doi: 10.1016/j.ecoleng.2014.03.045

Yun, D.H., and Y.T. Kim. 2018. Experimental study on settlement and scour characteristics of artificial reef with different reinforcement type and soil type. *Geotext. Geomembr.* 46(4):448–454. doi: 10.1016/j.geotexmem.2018.04.005

Zheng, X., T. Ji, S.M. Easa, and Y. Ye. 2018. Evaluating feasibility of using sea water curing for green artificial reef concrete. *Constr. Build. Mater.* 187:545–552. doi: 10.1016/j.conbuildmat.2018.07.140

4 Artificial Reefs in France
Current State-of-the-Art and Recent Innovative Projects

Sylvain Pioch, David de Monbrison, and François Simard

CONTENTS

ABSTRACT

The French artificial reefs history began in 1968, in Palavas-les-Flots, in the south of France, on the Mediterranean coast. From this time, more than 100,000 m³ of reefs were installed, mainly for fisheries enhancement and to protect sea-grounds against illegal trawling fisheries. Nevertheless, since 2010, a new development began, supported by European and French water agencies, to focus on biodiversity conservation. An important program of marine ecological engineering was initiated, with ecological restoration projects, and a nature-based solutions approach, including marine eco-design, using innovative technologies including: 3D-printing and biomimetism or eco-material. We will briefly present the past and the present, and try to draw the future of this useful coastal biodiversity management tool, dedicated to protecting and enhancing fish and their habitats, to sustain human activities in balance with nature's potential.

INTRODUCTION

In Europe, and France, fisheries management is closely linked with environmental enhancement. The ongoing target of creating Marine Protected Areas (MPAs) along 20% of the world's coastal areas by 2020 is progressing. It is important to note that France recently attained a level of 22% MPAs in January 2019. This effort in France is coupled with strong public investment to enhance the ecological functions in coastal areas. Ecological engineering projects are increasing and a number of them address using artificial reefs to restore benthic habitats and marine species targeted by fisheries. Current French policy advocates cost-efficiency while focusing on the benefits for biodiversity. These projects included assessing the performance of the artificial reef deployments in relation to ecological functions, and includes initial baseline investigations to determine the best approach over the long term. This chapter presents some of the most recent projects in France with regard to engineering artificial reefs for fisheries production, development of coastal activities (recreational, eco-mooring), and ecological restoration to offset human negative impacts.

MATERIALS AND METHODS

In Europe, most coastal countries have deployed artificial reefs to protect or enhance biodiversity and their fisheries. These environmental components are inextricably linked through their complex ecological functions to achieve a high quality of environmental health.

This examination includes two main periods of activities. This first activity period focused on enhancing/protecting fisheries while the second activity period targeted improving ecological function and a healthy environment (Pioch et al. 2010).

The period from the 1960s until the early 2010s (the Initiation Period) was dominated by the objectives to "enhance and protect fisheries." The reef project (Figure 4.1) was initially based on the European Union's fisheries policies which funded research to enhance target species and deter illegal trawling. This initial period of activities aligns well, particularly in recent years, to a period of development of Marine Protected Areas (MPAs). Projects conducted during this period also focused on optimizing costs for artificial reef deployments. This, however, was not fully realized because of insufficient basic information with regard to optimal artificial reef design, configuration, and distribution.

The end of this initiation period is characterized by the deployment of the largest artificial reef project (i.e., the PRADO Reef 2006; Figure 4.2) in the Mediterranean, confirming the progress of artificial reef development in France in accordance with an integrated approach. The reef complex was developed by the city of Marseille with the support of the BRL Ingénierie company and the academic community.

The project consisted of the installation of almost 28,000 m^3 of artificial reefs proximate to each other and to natural habitats, such as Posidonia meadows. Both the site and its governance were

FIGURE 4.1 Examples of French artificial reefs from the 1980s to 2010 (© BRLi).

FIGURE 4.2   Marseille Récif PRADO artificial reef complex (© BRLi).

identified in cooperation with local fishermen. Eight different types of artificial reef modules were deployed in sets between 20 m and 35 m apart to form the basis of these underwater "villages" to create the reef complex (Figure 4.2). The program continues to be monitored extensively with both passive and active acoustic techniques.

The second period (i.e., the Improving Period), followed the initiation period with an objective of "improving ecological function and a healthy environment" beginning in the early 2010s. Its beginning corresponds to the advent of massive pressure on the environment by fishing activities coupled with a notable reduction in biodiversity. Efforts during this period were supported by "green" financing associated with the European Union's water policies (Water Frame Directive, Marine Strategy Framework Directive) and the Habitat Fauna and Flora Directives (Natura 2000). Simultaneously, the gradual change in the public authorities' and society's concerns about ecological rehabilitation and restoration was growing to address ongoing biodiversity losses. The improving period led to the development of research projects with a new eco-engineering approach to improve the ecological performance of already deployed coastal structures. This effort included habitat restoration, advanced marine eco-friendly designs, artificial habitats, and artificial nurseries.

Since the first installation of artificial reefs in 1968 at Palavas-les-Flots in the south of France, the total volume of submerged artificial reef material has reached more than 100,000 m^3 (Tessier et al. 2015). Thus, many innovatively designed artificial reefs have only been installed in the last decade, with a total of around 50 sites defined as installed projects (Figure 4.3).

INNOVATIVE PERIOD INVOLVING ARTIFICIAL REEFS FOR FISHERIES PRODUCTION USING BIOMIMETISM

A third period of artificial reef development in France incorporated biomimetism. Four innovative biomimetic artificial reefs, designed to enhance fisheries, were installed in Ajaccio Bay (Corsica) by the Office of Environment of Corsica (OEC), LIB Industry Co., and Isula Service Co. These biomimetic artificial reefs are employed in a marine silicon skin, developed through research, applied into the mold during construction of the artificial reef module to mimic natural rock features (e.g., roughness, pits, and holes). An underwater photogrammetric (picture made with a three-dimensional image) system was developed to help mimic the natural rocky substratum where targeted species dwell (Figure 4.4). This provided an artificial reef with a microhabitat for prey species as well as their predators. This silicon mold technology can be used to create molds with rock features at the micrometric to the decimetric (~20 cm) scale.

INNOVATIVE ARTIFICIAL REEFS FOR COASTAL ACTIVITIES AND RESEARCH

The development of 3D-printed artificial reefs served to create marine nurseries and to enhance diving and other leisure activities, including education. These reefs were developed in the Mediterranean

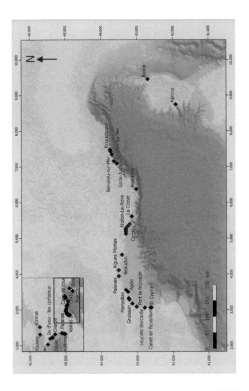
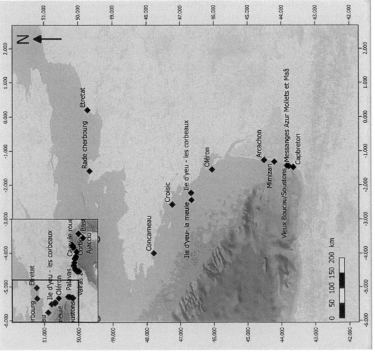

FIGURE 4.3 Map of the artificial reef sites in France—left square, in the Atlantic Ocean, and right square, in the Mediterranean Sea (© S. Pioch).

FIGURE 4.4 Photo of a biomimetic artificial reef modified for target fish species like sea bream, lobster, and stonefish (© S. Pioch).

Sea, mainly in France and Monaco. The nature of the materials used may vary (e.g., different types of concrete and sand) according to the projects, but nevertheless, the size of the modules is relatively small by volume. Two 3D-printed artificial reef projects are mentioned here. The first project was developed in 2018 off Monaco, with a module (Figure 4.5) designed to serve as a nursery, immersed at a depth of 28 m in the Larvotto Reserve in Monaco by the University of Nice and the Boskalis Company. The second project was conducted in 2019 off the city of Agde, built by XTREE Co. and Seaboost Co., and included 32 modules of 1.2 tons each immersed in shallow water (i.e., less than 10 m in depth).

Confronted with continuing changes in regulations and to better protect sensitive habitats, local and national authorities decided to promote organized mooring areas for small and large leisure

FIGURE 4.5 Photo of 3D-printed artificial reef off Monaco (© AMPN).

FIGURE 4.6 Illustration of new eco-mooring device (© S. Pioch).

vessels (Figure 4.6) along the French coast (Pioch et al. 2018). Several of these eco-mooring projects have been developed, employing innovative materials (Figure 4.7) with a functional approach to maximize protection of juvenile lobster shelters, thus creating corridors between the moorings and the natural environment.

These corridors were thought to increase local connectivity between the artificial-reef mooring area and the natural rocky area (coral reef) in the bay. Currently, in Deshaies, and soon afterward in Bouillante Bay (Guadeloupe), these ecological anchorages also serving as artificial reefs were

FIGURE 4.7 Photo of artificial mangrove belt and lobster cave in concrete, constituting the anchor for the eco-mooring system (© S. Pioch).

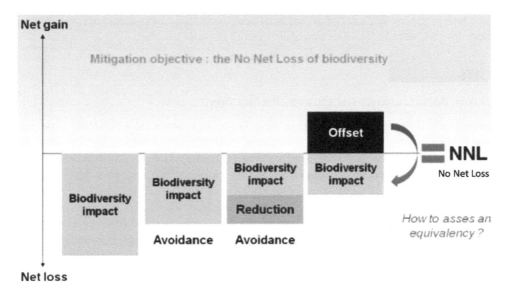

FIGURE 4.8   The Avoid–Reduce–Offset sequence under the Environmental Impact Assessment regulation, with the target of "No Net Loss" for compensatory mitigation (© S. Pioch).

designed for the French public harbor manager SEMSAMAR (Société d'économie mixte de Saint-Martin). The moorings were modified to accommodate the juvenile life stages of coral reef fishes. More specifically, they were designed to have features (e.g., size, depth, current orientation, color, and materials) to enhance the life history features of fishes targeted by fishermen (e.g., juvenile lobsters and snappers). The overall goal was to mimic natural habitats and protect juveniles against the invasive lionfish (*Pteroïs volitans*) imported by humans from the Indo-Pacific Ocean.

The primary purpose of the mooring was to facilitate yachting and boating by offering a safe mooring system. Secondarily, however, the moorings enhanced lobster and snapper juveniles and other locally targeted fishes with an eco-designed structure (Figure 4.7 and 4.8). The concrete base of the mooring was eco-designed (i.e., high roughness, pits, and caves mimicking natural hard substrate) with an additional artificial mangrove-like rim made of marine PEHD (polyethylene high-density) foam designed to accommodate juvenile snappers (Pioch and Leocadie 2017). At the top of the mooring base, a lobster cave was formed in concrete to allow inhabitation by cryptic fauna.

INNOVATIVE ARTIFICIAL REEFS FOR ECOLOGICAL RESTORATION AND MITIGATION

Artificial reefs for the purpose of facilitating ecological restoration continue to be developed. Several small projects are currently being developed to enhance biodiversity in ports and to improve coastal ecological infrastructure and performance through the development of ecological corridors. The incorporation of ecological corridors in the design are based on observations from natural seaweed transplantations (*Cystoseira amentacea*). These corridors can be incorporated into the design and placement of artificial reefs to enhance the use of artificial reefs in habitat damage mitigation.

Additionally, some artificial reef developments, supported by public finances, were deployed to investigate the filtering effect of mussels (*Mytilus galloprovincialis*) and other sessile biota located in front of urban outfall. The objective is to understand how artificial reefs can help restore ecological functions in polluted water areas by fixing and recycling nutriments from urban water (domestic water).

For instance, the "Rexcor" project conducted in Marseille during 2018, in the Parc National des Calanques (PNC - a marine protected area) examined the effect of the presence of artificial reefs on the water filtration of the city's sewage effluent located in the Bay of Cortiou (Marseille city coast). The reef surfaces should be able to provide substrate for the attachment of sessile, filter-feeding organisms. A total of 36 artificial reefs, composed of concrete and metal with three different designs, were installed to assess their utility in water filtration.

RESULTS

INNOVATIVE ARTIFICIAL REEFS FOR FISHERIES PRODUCTION

The biomimetic-designed artificial reefs in Ajaccio Bay have attracted 80% of the anticipated targeted fish species 1.5 years after their installation. Amberjack, sea bream, and stonefish were the first fishes to colonize the reef, followed by red sea bream and lobsters. Fishermen and the OEC (Office de l'Environnement de la Corse) are planning a larger reef complex for deployment in 2021 using copies of different kinds of natural rocks to provide shelter for juveniles.

The Récif Prado artificial reef complex was monitored and the data indicated its effectiveness in increasing fish biomass, presumably through production mechanisms. After 7 years of monitoring, the biomass was four times greater than prior to reef deployment, confirming its capacity to produce diversity and fishery resources for fishermen (Marseille City, personal communication with the executive officer of the artificial reef program Ms Cecilia Medioni). In the first year, it was an attractive area for fishes (adults in the first year were not produced from the artificial reef), then larvae and juveniles were observed, as well as spawning and breeding functions (end of the second year). Since the third year, evidence of species growing inside the area confirms the productivity effect (Cresson et al. 2014) e.g., *Octopus vulgaris, Loligo vulgaris, Sciaena umbra, Scorpaena porcus* and *Scorpaena scrofa* and *Diplodus sargus* and *Diplodus vulgaris*.

INNOVATIVE ARTIFICIAL REEFS FOR COASTAL ACTIVITIES

The results from monitoring the "eco-mooring" in Guadeloupe (Bouchon et al. 2017) are of interest. Indeed, one year after deployment, 58 benthic invertebrate species (including nine coral species and 43 fish species were recorded along with several lobster juveniles and four snapper species (i.e., species targeted by fishermen). Surprisingly, the proximate natural coral area had 45% fewer fish species than the eco-mooring habitat (i.e., 74 benthic invertebrate species and 25 fish species recorded in the adjacent coral area).

INNOVATIVE ARTIFICIAL REEFS FOR ECOLOGICAL RESTORATION AND MITIGATION

The XReef project is too recent for ecological results to be realized with sufficient monitoring time to allow evaluation of any ecological results. Regarding initial observations at the XReefs, colonization appears to be good while a few species like squid and red mullet have been observed (Agde city MPA, Director of the project, personal communication with Mr Renaud Dupuy de la Grandrive). It is not appropriate to speculate on the effects on ecological restoration and mitigation for this kind of artificial reef at this time.

DISCUSSION

Recently, artificial reefs in France and in Europe are being used less to enhance fisheries than to restore biodiversity. Ecological restoration forms the basis for enhancing the development of all marine species (including species of fisheries interest) as well as addressing the challenges offered by the influence of climate change or human pressure activities on the marine environment.

In France, the future development of artificial reef projects is mainly related to:

- An expansion of artificial reefs related to fisheries, to include biomimetism with a three-dimensional copy of natural habitats; mimicking natural habitats to restore spawning, breeding, sheltering, or feeding functions

- 3D-printed artificial reefs with specific focus on incorporating the relationship between reef complexity and materials with species protection and sustainability
- Development of new programs related to eco-designed infrastructures, ecological mooring coupling the effects of an artificial reef and the anchoring function, for large vessels (cruisers, yachts, and superyachts)
- Extension of application to nature-based solution systems for beach and coast anti-erosion: seaweed belts and oyster reefs on artificial reefs
- Development of new artificial reef programs to the Avoid–Reduce–Offset sequence associated with the evolution of Environmental Impact Assessment (EIA) policies and the target of No Net Loss (NNL) of biodiversity after any human construction project in the sea.

Ongoing programs to restore ecological functions are funded by the European Community and France, mainly under the policy of "water quality" programs and Natura 2000 regulations (i.e., the European legislation for natural habitat management), along with nature-based solutions, and other active and passive restoration projects. Therefore, environmental monitoring in the marine environment is now focusing on determining the gains and losses of ecological functions (and not only the classic attributes of functions such as abundance or species richness) (Figure 4.8).

For instance, the "MERCI-Cor" method (Pioch et al. 2017) helps in this evaluation by scoring the level of ecological function through ecological indicators (habitat, landscape, species, and water quality). This method is established to assess a compensatory mitigation measures to offset human impact, and obtain no net loss of biodiversity, targeted by the new French environmental law since 2016 (Biodiversity Law n° 2016-1087 of August 8, 2016). The method represents a key step in addressing Avoid–Reduce–Offset objectives, to maintain biodiversity as well as human activities. Recently, this progression in thinking led Monaco to incorporate within its latest town extension at sea project (i.e., the "Anse du Portier" project) various eco-designed elements to reduce its ecological impact, and losses of biodiversity. The scientific committee mobilized for this project recommended various initiatives, based on all these evolutions, for an eco-designed approach taking in account coastal ecological functions and deep-sea reef connectivity (ecological landscape approach).

To conclude, artificial reefs projects and research in France today are focused on ecological restoration targets as well as offsetting negative human impacts using eco-designed methods (Pioch et al. 2018) to "kill two birds with one stone." The future is finding and enhancing co-benefits for the natural environment and human activities in the marine area.

ACKNOWLEDGMENTS

The authors thank Shinya Otake and Yasushi Ito for their help and insights. We also thank l'Office de l'environnement de la Corse, Institute Mines-Télécom Alès, Jessica Salaun, the SEMSAMAR, and the cities of Marseille and Agde for their support and research projects.

REFERENCES

Bouchon, C., Y. Bouchon, S. De Lavigne, and S. Cordonier. 2017. *Suivi des communautés marines ZMEL Ecorécifs de Deshaies*. Université des Antilles, Borea et CAC.

Cresson, P., S. Ruitton, and M. Harmelin. 2014. Artificial reefs do increase secondary biomass production: Mechanisms evidenced by stable isotopes. *Marine Ecology – Progress Series* 509:15–26.

Pioch, S., and A. Leocadie. 2017. *Overview on Eco-Moorings Facilities: Commented Bibliography*. ICRI. https://www.icriforum.org/sites/default/files/OVERVIEW%20of%20eco-mooring-light.pdf (accessed March 3, 2020).

Pioch, S., M. Pinault, A. Brathwaite, A. Méchin, and N. Pascal. 2017. *Methodology for Scaling Mitigation and Compensatory Measures in Tropical Marine Ecosystems: MERCI-cor*. IFRECOR Handbook. https://www.icriforum.org/sites/default/files/HandBook%202%20-%20V5.pdf (accessed March 3, 2020).

Pioch, S., J.C. Raynal, and G. Lasserre. 2010. The artificial habitat, an evolutionary strategic tool for inte-
    grated coastal area management. In: *Global Change: Mankind-Marine Environment Interactions*, ed.
    H.J. Ceccaldi. Springer: Dordrecht, 129–134.

Pioch, S., G. Relini, J.C. Souche, M.J.F. Stive, D. De Monbrison, S. Nassif, F. Simard, P. Saussol, R. Spieler,
    and K. Kilfoyle. 2018. Enhancing eco-engineering of coastal infrastructure with eco-design: Moving
    from mitigation to integration. *Ecological Engineering* 120:574–584.

Tessier, A., P. Francour, E. Charbonnel, N. Dalias, P. Bodilis, W. Seaman, and P. Lenfant. 2015. Assessment
    of French artificial reefs: Due to limitations of research, trends may be misleading. *Hydrobiologia*
    753(1):1–29.

5 Development and Utilization of Artificial Reefs in Korea

Lee Moon Ock, Oh Tae Geon, Baek Sang Ho, and Kim Jong Kyu

CONTENTS

ABSTRACT

This chapter discusses the current situation on utilization and development of artificial reefs in Korea. Artificial reefs have been introduced in Korea since the first project officially started in 1971. Between 1998 to 2015, comprehensive projects have been implemented to improve the fishing industry. These projects include the release of seedling fish, and the construction of marine ranches and sea-forests using technologies of design and deployment of artificial reefs. Subsequent to the initiation of these projects, fish catches have increased to roughly double the catch before the deployment of artificial reefs. In addition, fishermen's earnings have also improved. Accordingly, local governments and communities are closely collaborating to rehabilitate the fishing industry by improving the marine environments, conducting complementary planting of algae, organizing self-governing fishing communities, and conducting post-management activities associated with marine ranching. Notably, in recent years, fishing industries in nearly all parts of the world have been challenged due to climate change, a rise in sea level, and coastal pollution. Consequently, we must share our knowledge of artificial reefs and collaborate in order to overcome the current crisis facing the fishing industry.

INTRODUCTION

Artificial reefs are structures constructed in aquatic environments to attract and concentrate aquatic organisms and to potentially improve and rehabilitate coastal ecosystems (Pickering 1996, Pickering et al. 1997, Pickering et al. 1998). Seaman and Jensen (2000) also defined an artificial reef as "one or more objects of natural or human origin deployed purposefully on the seafloor to influence physical, biological, or socioeconomic processes related to living marine resources." Today, artificial reefs are being constructed in coastal areas around the world for various purposes: e.g., to improve fishery production, preserve biodiversity, protect habitats, prevent illegal fishing, enhance recreational fishing, and facilitate tourism (Baine 2001, Bortone et al. 2011, Lee et al. 2018).

Artificial reefs are known to have been first used by humans around the 1640s in Japan (Ogawa 1968). Concomitantly, Koreans have been exchanging various cultural aspects of their fishing technologies with the Japanese during their long history. Thus, it is no wonder that there are many similar developments and utilization of artificial reefs between Korea and Japan. It would be difficult to deny that Japan is the state-of-the-art country in the world with regard to the technology concerning the development and utilization of artificial reefs. As a result, the development, research, and implementation of artificial reefs in Korea has been influenced by the use of artificial reefs in Japan. However, the application of artificial reefs under actual conditions at sea can be quite different, because artificial reefs have to be deployed in accordance with local socio-economic needs and geographical conditions. This study examines the current situations with regard to the utilization and development of artificial reefs, and how to use and manage them for the fishing industry in Korea.

MATERIALS AND METHODS

The data sources used here include periodicals, technical reports, business manuals, articles, and proceedings of symposiums and conferences that the Korea Fisheries Resources Agency (FIRA; http://www.fira.or.kr) conducted in relation to artificial reefs and marine ranch projects from 2010 to 2018. We also reviewed articles identified through the Science Citation Index (SCI) about artificial reefs for the last 20 years (1996 to 2017). Based on a review of these materials we elucidated the functional characteristics of artificial reefs, technical methods of their design, deployment, and management, and their socio-economic effects on the local community in Korea.

RESULTS AND DISCUSSION

HISTORY OF ARTIFICIAL REEF PROJECTS

Fisheries productivity in Korean coastal waters has decreased since the 1970s (as seen in Figure 5.1), due to various causes such as overfishing, pollution of coastal waters, and land reclamations. Moreover, the proclamation of a 200-mile Exclusive Economic Zone (EEZ) by many coastal nations has made access to deep-sea fisheries more difficult.

The history of artificial reef development in Korea is obscure, but documented artificial reef projects were started by the central government in 1971 (Kim 2004). In the meantime, great efforts have

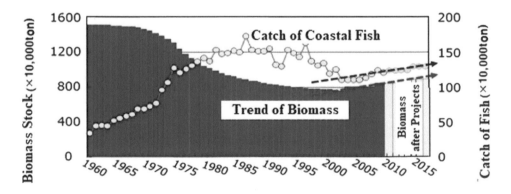

FIGURE 5.1 Performance of fisheries production in Korean coastal waters.

been expended to identify ways to utilize artificial reefs to aid in the recovery of fisheries resources. As a result, one of the strategies realized was to increase fish stocks by developing artificial reefs to facilitate spawning and improve nursery areas. Thus, 1.4 million units (or modules) of artificial reefs have been deployed nationally in 0.23 million ha of coastal habitat by 2015. Owing to this effort, the fish catch increased by 1.5–2.9 times compared to the reference site during the last 5 years (from 2013–2017). Likewise, large scale national projects began a seedling (i.e., juveniles of the target species) release of fish in 1986, established marine ranches in 1998, and constructed sea forests in 2009. In particular, the Korea Fisheries Resources Agency (FIRA) was founded in 2011 as a public corporation on the basis of a law called the Fisheries Resources Management Act. Subsequently, this agency took full charge of public projects involving the protection and promotion of fisheries resources, research, development, and the propagation of fishing ground management technologies. Accordingly, FIRA had constructed coastal marine ranches in roughly 20,000 ha at 36 sites by 2015, with plans to create coastal marine ranches at a total of 500 sites by 2030.

Domestically, various artificial and seaweed reefs have been developed and were included in these projects. In particular, sea forests were created with an area of 33,000 ha at 21 coastal sites to cope with the occurrence of isoyake (a whitening event) since 2009 (Japan Fisheries Agency 2007). The results of these activities are annually monitored through *in-situ* investigations using a multi-spectral scanner.

MATERIALS, SHAPES, AND DESIGN OF ARTIFICIAL REEFS

At present, Korea usually uses concrete and steel as materials for constructing artificial reefs. Additionally, fiber-reinforced plastic (FRP), ceramic, scrapped ships, plastic waste, oyster shells, biodegradable plastic, polyvinyl resin, slag, and yellow clay are sometimes used in the construction of these artificial reefs (Japan Large Size Artificial Reefs Association; http://www.nissyoukyou .com). Figure 5.2 indicates a process of construction and deployment for concrete reefs.

Common shapes of the artificial reef modules or units developed are square or cylindrical designs. Presently, a total of 87 different types of modules are now designated as acceptable as artificial reefs for deployment in the marine environment off Korea (Lee et al. 2017, FIRA 2019). Figure 5.3 depicts 28 different artificial reef modules. Some units are specifically designed for fish (Figure 5.3 a–h), shellfish and seaweed (Figure 5.3 i–w), fish and shellfish (Figure 5.3 x y), sea forests (Figure 5.3 z–aa), or shellfish and sea forests (Figure 5.3 bb). In particular, it is notable that a scrapped ship is generally used for fish (Figure 5.3 g), and ceramic (Figure 5.3 j) or yellow clay (Figure 5.3 v) are used as artificial reef materials for shellfish and seaweed, respectively.

In Korea, artificial reefs for fish are usually deployed at a depth of 15 to 70 m while artificial reefs for shellfish or seaweed are installed in coastal waters of less than 15 m in depth. Moreover, FIRA has established some regulations with regard to the selection of the areas suitable for deployments of artificial reefs. Thus, there is a need to determine an appropriate site in advance of an artificial reef deployment. In general, the pre-deployment survey area is minimally 8 ha for fish and 2 ha for seaweed or shellfish. Moreover, to determine whether or not the site is appropriate before the installation of artificial reefs, assessments are conducted for other environmental factors, such as DO, pH, currents, and topography.

When considering a design for artificial reefs, one should determine the intended target species, because each species may react differently to various features of an artificial reef. For example, Nakamura(1985) stated that Type A species prefer physical contact with artificial reefs while Type B species tend to swim around the artificial reefs. In addition, Type C species appear to prefer with artificial reefs that have void spaces. These designation types can be referred to as a species' affinity to artificial reefs. As a result, structural characteristics of the artificial reef modules and their arrangement can be designed in accordance with a target fish species and its life stages.

Artificial reef modules are not usually deployed singly but are most often deployed as part of a group (a set or in an array) of similar modules or with many different modules. For example, natural rocks are often deployed as artificial reef modules. Overall, the weight or density of artificial reefs

(1)   (2)   (3)

(4)   (5)   (6)

(7)   (8)   (9)

(10)

FIGURE 5.2 Manufacturing process of a concrete artificial reef and its deployment.

is important because these features are associated with the stability of artificial reefs in the open sea (FIRA 2015). Figure 5.4 depicts examples of artificial reef deployments.

Function of Artificial Reefs

Artificial reefs in Korea generally follow designs regulated by FIRA to satisfy physical, chemical, biological, engineering, social, and economic criteria. However, artificial reefs are mainly designed to fulfill functional objective from physical, ecological, and economical viewpoints. First, artificial reefs are often required to physically enhance a habitat to create upwelling, shadow zones, eddies, and seawater exchange, and to stabilize the substrate. Second, artificial reefs are required ecologically to provide nutrients, facilitate aggregation, enhance feeding, and offer shelter for fish. Third, artificial reefs must also be economically efficient to manufacture and maintain. Moreover, it is crucial for artificial reefs to be durable. In particular, artificial reefs should maintain their original function after deployment. Thus, artificial reefs should be structurally designed to avoid being entangled in nets or other fishing gear. Besides this, artificial reefs should not become deformed or harmful to marine organisms after deployment.

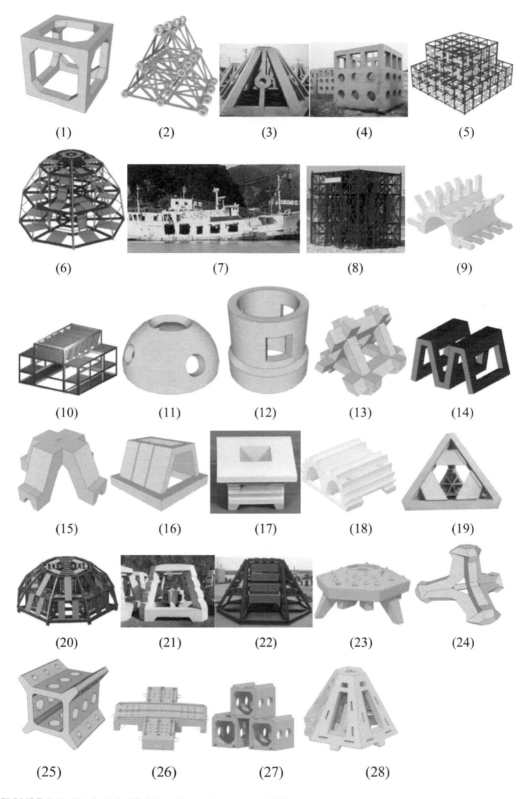

FIGURE 5.3 Typical Artificial reefs used at present in Korea.

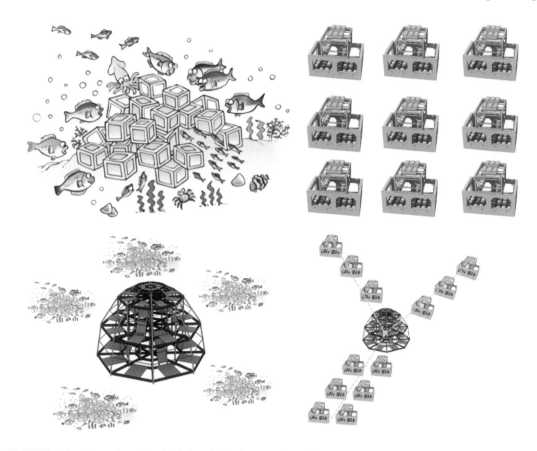

FIGURE 5.4   Examples of Artificial reefs' deployment in real seas.

MARINE RANCH PROJECTS

Background

Recently, the decrease in fisheries resources has become an important issue around the world especially in coastal waters. This large-scale decrease might be attributable to global climate change, overfishing, marine pollution, or unregulated coastal development. The establishment of the Exclusive Economic Zone may have also affected downscaling of pelagic or coastal fishing grounds. In particular, the ability of fishermen to earn their living has become more challenging than before the contract of Free Trade Agreement (FTA). Consequently, there is an urgency to institute measures to substantially increase fishermen's incomes through a revitalization of the fishing-village economy. Thus, marine ranch projects were implemented in Korea as one of several management systems to promote sustainable utilization of fisheries resources (FIRA 2016). Figure 5.5 depicts several usages of marine ranches that Korea presently manages and operates. The usage of a marine ranch is determined by the characteristics of the location, the nature of local businesses, and the kinds of resources targeted. Generally, marine ranches can be classified according to three uses: commercial fishing, recreation, and mixed (i.e., both commercial fishing and recreation) usage. Furthermore, there are various habitats that are used primarily for commercial fishing such as archipelagos, tidal flats, inner bays, and the open sea. Areas used primarily for recreational activities include underwater, tidal flats, and recreational fishing habitats. The locations of marine ranches are indicated in Figure 5.6, and 35 sites are already completed and are operating (green color), and 15 sites (red color) are underway.

Archipelago mode

Inner bay mode

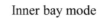

Open sea mode

Underwater experience mode

Tidal flat experience mode

Recreational fishing mode

Mixed mode

FIGURE 5.5   Several types of marine ranch in Korea.

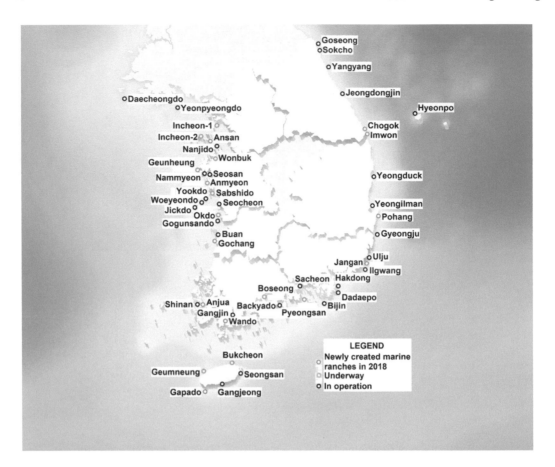

FIGURE 5.6   Location of a marine ranch.

Objectives and Functions of Marine Ranches

A marine ranch refers to an area that is artificially constructed in the marine environment to efficiently use fisheries resources under a systematic management plan to attain specific objectives, just like a ranch on land. Thus, a marine ranch fulfills a function as an area for the protection and growth of fisheries organisms. Moreover, it contributes to the sustainable production of fisheries resources, as well as the enhancement of tourism, so this has additional benefits for the economy of a fishing village. However, it should be realized that these objectives for the construction of a marine ranch can be only achieved with the implementation of technologies associated with the design and deployment of artificial reefs or their management.

Promotion Procedures for Marine Ranch Construction

When constructing a marine ranch in Korea, a local government should first propose an application to Ministry of Oceans and Fisheries (MOF) for a master plan. Then an execution plan is established if the master plan is approved as reasonable through *in-situ* site investigation or analysis of economic feasibility. Once the project site is approved by MOF, then substantial plans that involve the construction of the marine ranch with its follow-up management and evaluation of the ranch's performance are subject to the supervision of the FIRA or local government. Specifically, there is a committee called the Marine Ranch Self-Management Committee which is organized in accordance with tasks to select the species, and decide the quantity and specifications for a seedling release. Figure 5.7 depicts the process of a seedling release in marine ranch.

(1) Check a seedling condition

(2) Check the quantity of a seedling

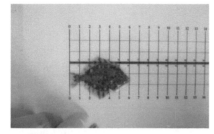
(3) Check a specification of a seedling

(4) A supervisor measure a seeedling

(5) Sampling for infection check

(6) Seal a sample

(7) Weigh a seedling

(8) Convey a seedling

(9) Release a seedling onboard

(10) Release a seedling at sea

FIGURE 5.7 Process of seedling release in the marine ranch.

The number of species managed using marine ranches in Korea is currently 65, including 50 marine species and 15 freshwater species. The marine species include: abalone, flounder, tiger puffer, sea urchin, filefish, rockfish, sea bream, scorpion fish, black porgy, sole, sea bass, cod, scallops, and clams. Inland freshwater species include: king crab, carp, catfish, mandarin fish, eel, terrapin, sweetfish, marsh snail, loach, and corbicula. However, the species being raised in the marine ranch are different from one another depending on the seas i.e., in the east, west, south, or Jeju. Additionally, there is a difference in the size of individuals of each species, as well as the seasons when fish from other areas are released in the marine ranch.

POLICY FOR FISHERIES REHABILITATION

FIRA promotes several rehabilitation projects for the fishing industry as part of its government activities. For example, FIRA conducts projects to improve marine habitats, which include the extermination of seaweed-eating organisms (e.g., sea urchin and starfish), release of plant seaweed/algae, restoration of substrate for seaweed/algae, and sediment improvement. Figure 5.8 shows some activities to improve the marine environment.

FIRA also operates some facilities so that people can enjoy the experience of being in the marine environment, and these include marine vacation areas and recreational fishing grounds, as shown in Figure 5.9. With all of these projects, the fishing industry should become rehabilitated in the near-future, in terms of making the seas more valuable and healthier than ever before.

(1) Cleanup   (2) Exterminate a starfish

(3) Transplant seaweeds   (4) Plow for better tidal flats

(5) Create newly oyster farms   (6) Scatter parent shellfish

FIGURE 5.8   Some activities for rehabilitation of marine environment.

SURVEYS TO DETERMINE EFFECTIVENESS AND THE POST-MANAGEMENT OF MARINE RANCHING

After a marine ranch has been constructed, it is necessary to verify whether its function is properly performing or not. Thus, various field surveys are conducted in the marine ranch area as follows (see Figure 5.10):

(1) Collect samples of fish and invertebrate using a gill net or quadrat sampling.
(2) Investigate the standing stock of pelagic fish using an acoustic fish-finder.
(3) Investigate the benthic biota and its density using a haul net.
(4) Investigate periphyton and algal succession by means of scuba-diving or quadrat sampling.
(5) Survey the environmental conditions around the marine ranch by scuba-diving, side-scan sonar and multi-beam echo-sounder.
(6) Collect feedback from fishermen and analyze socio-economic effectiveness.

A marine ranch could become functionally degraded with time, even if natural conditions are maintained. Thus, there is a need for post-management activities to systematically maintain the sustainable function of the marine ranch. Figure 5.11 indicates the utility of establishing post-management activities. A lack of post-management actions could result in a reduction in the effective function of a marine ranch.

To accomplish these activities, FIRA is closely associated with local government and communities to maximize the function of marine ranches. FIRA, in concert with local governments and communities, conducts the following projects, as indicated in Figure 5.12:

(1) Improve marine environments to provide a marine life with a better environment.
(2) Repair and reinforce degraded functions of artificial reefs through conducting research on the effective function of the structure.
(3) Release seedling fish.
(4) Designate fishing resource management waters which contribute to the efficient management and sustainable use of marine resources via the marine ranch.
(5) Organize fishing communities to become self-governing in the operation of the marine ranch.

(1) Marine rest facility

(2) Marine fishing hole (floating type)

(3) Marine fishing hole (fixed type)

(4) Underwater experience facility

FIGURE 5.9 Marine facilities for rest, fishing, and sports.

(1) Collect a sample of fish and invertebrate by a gill net and a hatch

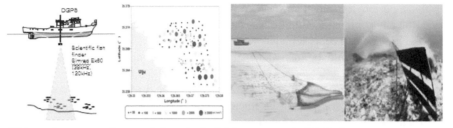

(2) Investigate a standing stock for pelagic fish by an acoustic fish finder

(3) Investigate a biota of benthos and periphyton by a scuba diving and a quadrat

(4) Investigate a facility's condition by a side scan sonar and multi beam echo sounder

(5) Analyze a socioeconomic effectiveness (B/C ratio) by a collection of a feedback from fishermen

FIGURE 5.10   Effectiveness survey of marine ranch.

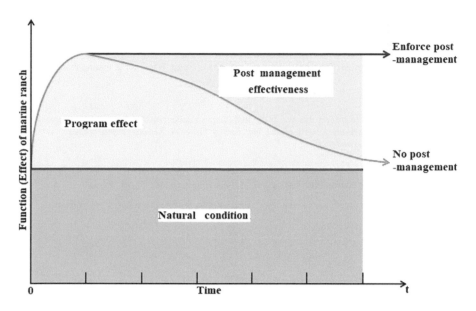

FIGURE 5.11 Concept of post-management after construction of marine ranch.

[Appended map]

Location and its area of Fishing Resource Management Water

1. Location : Gilcheon-ri, Weollae-ri, and Imryang-ri, Jangan, Gijang-gun, Busan
(The coordinates of each point, i.e. (a), (b), (c), (d), and (e) are expressed as WGS84 system)

2. Section : Water area enclosed by the line that successively connects from (a) to (e).

3. Area : 1,700,000m²

  (a) 35°18.9′N, 129°16.1′E

  (b) 35°19.4′N, 129°16.9′E

  (c) 35°18.9′N, 129°16.9′E

  (d) 35°18.7′N, 129°17.8′E

  (e) 35°18.4′N, 129°17.6′E

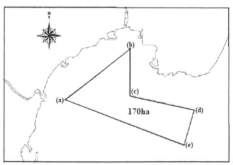

FIGURE 5.12 A variety of projects for post-management.

Busan Official Notice No. 2016-1592

Notice of the designation to Fishing Resource Management Water for a marine ranch in Jangan Area

Here is a notice to inform that we designated Jangan marine ranch as the fishing resource management water on a basis of the law named Fisheries Resources Management Act (Article 48, No. 1) as follows:

1. Name : Jangan marine ranch fishing resource management water, Gijang, Busan

2. Location : The location is to be water area enclosed by the line that successively connects

from (a) to (e) as follows:

    (a) 35°18.9′N, 129°16.1′E

    (b) 35°19.4′N, 129°16.9′E

    (c) 35°18.9′N, 129°16.9′E

    (d) 35°18.7′N, 129°17.8′E

    (e) 35°18.4′N, 129°17.6′E

3. Area : 170 ha, inside area of the line connected by from (a) to (e)

4. Period : 5year

5. Manager : Governor

6. Acceptable activities :

    (a) Collection of aquatic organisms or installation of structures for investigation and research

    (b) Installation of structures for fisheries resource creation, a seedling release, cleanup, extermination of harmful organisms

    (c) Access of a ship for management

    (d) Angling or fishing by hands

    (e) Recreational fishing and skin scuba diving in a permitted area

    (f) Others to be approved by the committee of administration or mayor of Busan

FIGURE 5.12   Continued.

FIGURE 5.12 Continued.

CONCLUSIONS

Relative to the development and utilization of artificial reefs in Korea, here we reviewed technical reports, business manuals and guidelines, and proceedings of the conferences that were mainly published by MOF/FIRA during the past decades.

Korea officially initiated artificial reef projects in 1971 with a goal not only to enhance a fishing industry, but to increase fishermen's earnings. As a result, the fish catch has doubled during the last five years and, accordingly, fishermen's income apparently increased to some extent. However, questions remain as to how much the fisheries resources actually increased and if the ecosystem's health improved. Also, we do not know exactly how those artificial reef projects contributed to the increases in biodiversity and biomass. Nevertheless, the indications are the present improving conditions are the result of these activities.

Moreover, Korea conducted a seedling release project of fish in 1986 following a marine ranch project in 1998 and a sea forest creation project in 2009. The objectives for these projects were to secure a stable fisheries production and also to mitigate isoyake (a whitening event). The result was that many fishing villages regained their economic vigor through an increase in eco-tourism.

Currently, the fishing industry is being challenged due to global warming, a rise in sea levels, and increases in coastal pollutions. Additionally, the sustainability of fish stocks is threatened by man-made wastes such as micro-plastics. Therefore, future management efforts will need to share technologies related to the development and utilization of artificial reefs. Close collaboration of all concerned is needed in order to overcome the current crisis in the Korean fishing industry.

REFERENCES

Baine, M. 2001. Artificial reefs: A review of their design, application, management and performance. *Ocean & Coast Management* 44(3–4):241–259.

Bortone, S.A., F.P. Brandini, G. Fabi, and S. Otake. 2011. *Artificial Reefs in Fisheries Management*. CRC Press/Taylor & Francis, Boca Raton, FL.

Japan Fisheries Agency (JFA). 2007. *Guideline for the Countermeasures on Isoyake*. Ministry of Agriculture, Forestry and Fisheries, 1–208 (in Japanese).

Japan Large Size Artificial Reefs Association. http://www.nissyoukyou.com (accessed August 3, 2015).

Kim, H.S. 2004. Deterioration characteristics of reinforced concrete reefs submerged in seawater. PhD dissertation. Pukyong National University, Pusan, 1–167 (in Korean).

Korea Fisheries Resources Agency (FIRA). 2016. *Guideline for Marine Ranch Creation Process*. Ministry of Oceans and Fisheries, 1–192 (in Korean).

Korea Fisheries Resources Agency (FIRA). 2019. *Korean Artificial Reefs Information*. Ministry of Oceans and Fisheries, 1–96 (in Korean).

Lee, M.O., S. Otake, S.H. Back, and J.K. Kim. 2017. Development and utilization of artificial reefs (ARs) in Korea and Japan. *Fisheries Engineering* 54(1):23–30 (in Japanese).

Lee, M.O., S. Otake, and J.K. Kim. 2018. Transition of artificial reefs (ARs) and its prospects. *Ocean and Coastal Management* 154:55–65.

Nakamura, M. 1985. Evolution of artificial fishing reef concepts in Japan. *Bulletin of Marine Science* 37:271–278.

Ogawa, Y. 1968. Artificial reef and fish. *Propagation of Fisheries* 7:1–21 (in Japanese).

Pickering, H. 1996. Artificial reefs of bulk waste materials: A scientific and legal review of the suitability of using the cement stabilised by-products of coal-fired power stations. *Marine Policy* 30(6):483–497.

Pickering, H., and D. Whitmarsh. 1997. Artificial reefs and fisheries exploitation: A review of the 'attraction versus production' debate, the influence of design and its significance for policy. *Fisheries Research* 31(1–2):39–59.

Pickering, H., D. Whitmarsh, and A. Jensen. 1998. Artificial reefs as a tool to aid rehabilitation of coastal ecosystems: Investigating the potential. *Marine Pollution Bulletin* 37(8–12):505–514.

Seaman, W., Jr. and A.C. Jensen. 2000. Purposes and practices of artificial reef evaluation. In: *Artificial Reef Evaluation*, ed. W. Seaman. CRC Press/Taylor & Francis, Boca Raton, FL, 1–19.

6 The Status of Artisanal Fish Aggregating Devices in Southeast Asia

Jarina Mohd Jani

CONTENTS

ABSTRACT

In the Southeast Asia region, various versions of artisanal fish aggregating devices (FADs) have been, and continue to be, part of the traditional local fishing practices. The use of artisanal FADs that were recorded in colonial documents did not cease with the introduction of more modern artificial reefs four decades ago. However, while the development and progress of artificial reefs in this region are well documented, there is a dearth of information on the artisanal FADs. This chapter provides a review on the status of artisanal FADs in this socio-culturally diverse marine region based on information gathered from official reports and research papers, as well as alternative sources such as web-based information shared by the public. The chapter highlights artisanal FADs as a regional commonality that has an enduring presence, particularly in countries where modern artificial reefs are less extensively used. It is notable that these artisanal FADs are given varied levels of attention in fishery research and management. Moreover, many of these older, relatively primitive devices have been modified to enhance their relevance in the contemporary fishery context. Lessons learned from these findings are considered in proposing the contributions that research in artisanal fish aggregating devices could make towards the sustainable management of the fisheries resources at local and regional levels.

INTRODUCTION

A fish aggregating device (FAD) is "a permanent, semi-permanent or temporary structure or device made from any material and used to lure fish" (FAO 2019). It is an artificial habitat that has long been used by fishers from all over the world to facilitate fishing operations (Dempster and Taquet 2004). Similarly, in Southeast Asia, coastal fishers have long used artisanal FADs and benthic artificial reefs. The use of FADs in Malaysia, Indonesia and the Philippines has been well-documented in artificial reefs-related literature (e.g., Bergstrom 1983, Delmendo 1990, Chou 1997). Created by fishers using various materials available to them (such as coconut or other palm fronds, bamboo, and leafy tree branches, etc.), these devices intrigued early researchers. In Malaysia, for example, Brown (1935) and Parry (1954) reported the ingenuity of local fishers along the East Coast of the Peninsular in using the *unjang* (i.e., fish aggregating devices; spelled according to the local Terengganu dialect. The standard Malay spelling is *unjam*, and on the western coast of the peninsula, it was also called *tuas*) to facilitate their fishing operations, while Firth (1975) highlighted the local fisheries management and trade practices associated with it as well. Dutch scientists mentioned that similar contraptions called *rumpon* were used in the West Sumatran and the Sulu-Sulawesi marine regions (Yusfiandayani 2013). In the Philippines, they are known as *payao*. Since 2000, additional information related to artisanal FADs in the Gulf of Thailand has been recorded. *Sung* (a fish aggregating device similar to *payao*) were introduced in the 1970s to facilitate Chinese-style, purse seine fishery that targeted pelagic species (Noranarttragoon et al. 2013). Less is known about artisanal FADs in Cambodia where a traditional brush park fishery known as *samrah* is practiced in the Great Tonle Sap (Ho 1999). In Vietnam, artisanal FADs were used by the Sampan people in Tam Giang Lagoon (Brzeski and Newkirk 2000). References to such devices, however, did not appear in any of the comprehensive records on artisanal fishing that had been reported from Singapore (Parry 1954) or Brunei (Beales 1982). Another country where no information on artisanal FADs was found, either formally or informally, is Laos, a landlocked country.

Artisanal FADs have continued to be deployed alongside modern artificial reefs that have been introduced since the 1970s, when the latter were widely introduced in the greater Southeast Asian region, as discussed by Delmendo (1990) and Ebbers (2003). However, by the 1990s, artificial reefs had become important in state fisheries management in some countries (Chou 1997). In fact, due to their unique connection to artisanal FADs, artificial reefs were lauded as important in the successful implementation of community-based fisheries management in Southeast Asia. These reefs were considered a solution toward mitigating the depleting resource management issues in the region (Yahaya 1993). As part of national fisheries management plans in many Southeast Asian countries, the status of modern artificial reefs in the region continues to be well documented (see Ali et al. 2011) as part of many countries' fisheries enhancement programs (Siriraksophon and Sayan 2016). But, to date, no regional review has been conducted exclusively on the artisanal versions of artificial reefs. Perhaps less attention has been given to artisanal FADs due to their relatively unsophisticated nature, even though their importance has not lessened since the modern development and deployment of artificial reefs (Supongpan 2006). This lack of critical review may be particularly valid in the less developed parts of the vast marine areas of this region.

The innovative features of modified artisanal FADs are discussed before summarizing the perspectives learned regarding their deployment in various situations. This chapter concludes by suggesting a way toward capitalizing on the changes in artisanal FADs to ensure sustainable fisheries in the region.

MATERIALS AND METHODS

This chapter addresses this knowledge gap by presenting an overview of the diverse types of artisanal FADS that have been recorded in the Southeast Asian region. It is based on relevant information on fish aggregating devices gathered from online-sourced official reports and public materials from each country in Southeast Asia. Also included in this review is information collected during field observations and informal interviews with small-scale fishers at various sites in the region. Additionally, country-specific reports and scientific articles related to fisheries management and artificial reefs, that often provided a hierarchical perspective on artisanal FAD development, were examined. Moreover, other information gathered from informal sources, such as blogs and private websites, provided a much-needed basic perspective. Together, these sources provided an overall perspective of the current status of artisanal FADs in Southeast Asia. In total, these sources enabled the author to compile a background on the history and development of artisanal FADs for each country in the region. Additionally, this information enabled a perspective of their persistent and continued use evaluated using a set of criteria applied to the information found using multiple Google™ search engines such as Google Search, Google Image, and Google Scholar.

RESULTS AND DISCUSSION

ARTISANAL FADs IN SOUTH EAST ASIA: AN ENDURING PRACTICE

In addition to countries where the presence of artisanal FADs has been recorded (i.e., prior to 2000) in the relevant literature on artificial reefs, recent use of artisanal FADs was recorded off Timor-Leste and Myanmar (see Table 6.1). For the former location, the discovery of new records of artificial reefs, mainly in the form of artisanal FADs similar to those found in Indonesia, was enabled by a change in political conditions as Timor-Leste gained independence only two decades ago. As for Myanmar, which remains under military rule, there is still little official information on this topic. However, advances in, and availability of, internet technology has made it possible to obtain and share some information both within and extramurally from Myanmar, where according to the Myanmarese Department of Fisheries, there is no need to develop artificial reefs due to the healthy condition of its marine ecosystem (FOA 2006). Thanks to an underwater image taken for a marine NGO that works with the Moken sea-nomads (see www.projectmoken.com), we are able to catch a rare glimpse of a unique, now illegal, artisanal FAD used by the Moken tribe to facilitate fishing while skin-diving (see Figure 6.1).

TABLE 6.1
Types of Artisanal FADs in Southeast Asian Countries and an Evaluation of Their Continued Use

Country	Recorded presence	Types of artisanal FADs recorded	Continued use*
Vietnam	Yes	*Payao* (recent introduction in association with purse seine fishery)	Moderate
Cambodia	No	*Samrah* (Brush parks used in Tonic Sap Lake)	Nil
Laos	No	–	Nil
Myanmar	Yes	*Moken* FAD: semi-submerged floating coconut leaves	Low
Thailand	Yes	*Sung*: semi-submerged coconut fronds with bamboo contraptions (an introduced version of *payao* used in offshore fishery)	High
Malaysia	Yes	*Unjang* or *tuas*: semi-submerged coconut fronds with leafy branches	Moderate
Indonesia	Yes	*Rumpon, rakit, tendak, pocong*: semi-submerged coconut or nypa fronds with leafy branches,	High
Philippines	Yes	*Payao*: semi-submerged coconut fronds with bamboo contraptions	High
Singapore	No	–	Nil
Brunei	No	–	Nil
Timor-Leste	Yes	*Rumpon* and floating *payao*-like bamboo structure	Moderate

*Criteria used for evaluation based on material found via Google Search Engine, including Google Scholar and Google Image:

High: Relevant information is sufficiently available (>50% of material found); sufficient formal (researchers and managers) and informal (users) content.

Moderate: Relevant information is available (20–50%); insufficient formal (researchers and managers) or informal (users) content.

Low: Relevant information is hardly available (1–20% of material found); insufficient formal (researchers and managers) and informal (users) content.

Nil: No relevant information is available (0% of material found).

FIGURE 6.1 Artisanal FAD called "Bamboo-Island" used by the nomadic Moken tribe in Myanmar. (Photo credit: Sofie Olsen for Project Moken.)

In countries where there are no records of artisanal FADs, their continued absence may be due to the lack of either socio-economic or ecological conditions. In the most developed countries of the region (e.g., Singapore and Brunei) where traditional fisheries are not socio-economically significant, there is no reason to promote the deployment of artisanal FADs. Development of other types of artificial reefs, however, continues in these countries, but mostly for conservation purposes (Supongpan 2006). As for Laos and Cambodia, artisanal FADs appear unsuitable for the predominantly inland fisheries in these countries. In addition, there is perhaps little need for such deployments in the Cambodian marine capture fishery, due to the relative abundance of Cambodia's marine coastal resources (Srean 2018).

Considering the recent prohibition of the "Banana-Island" FAD method by the Myanmar authorities, the continued use of these awe-inspiring primitive artisanal FADs is uncertain. In Timor-Leste, however, traditional devices are innovated into modified artisanal FADs to better exploit the country's off-shore fishery thanks to encouragement from the state's fishery development agency. Here, where artisanal FAD usage is considered as moderate, they are deployed near the coast to assist line-fishers. Additionally, the use of the more *payao*-like FADs is promoted to help the East Timorese local fishers take advantage of the premium tuna fishery located farther offshore than their usual fishing grounds (Beverly et al. 2012).

In countries where artisanal FADs have previously been recorded, the levels of use differ. Usage is rated as moderate for Malaysia, because the attention given to artisanal FADs is diminished in comparison to the more prominent, modern versions of FADs. There appears to be only a single researcher continuing to work on artisanal FAD research in Malaysia (see Mohd Jani 2015 and Mohd Jani et al. 2018). However, information regarding the construction of these devices, specifically the *unjang*, is popularly shared by users, particularly by anglers who use FADs created by local fishers for whom the *unjang* remains an important element in their coastal fishing operations. Artisanal FAD use is also moderate in Vietnam, where the fishers employ only *payao*-like FADs in tuna fishery (Vinh 2001).

As for the remaining three countries (Indonesia, the Philippines, and Thailand), the current usage of artisanal FADs is rated high. In Indonesia, artisanal FADs have gained the keen interest of fisheries researchers and managers. This is apparent from the increasing number of publications on the subject by these individuals, particularly with regard to their published studies on the tuna fishery (see Yusfidayani 2013, Wiadnya et al. 2018, Natsir et al. 2018). The published information by these researchers matches the abundant online information currently available. Much of this new information comes from fishers in this vast archipelago of Southeast Asia that now has widespread, affordable 4G cellular communication coverage. However, there is concern that the *payao*-like FADs, used in Indonesian waters to catch tuna, are reportedly deployed by foreign fishers. Most of these fishers are from the Philippines where the *payao* originated. Indeed, as early as the 1980s, the *payao* had been deployed to facilitate fishing throughout the Asian Pacific region (Aprieto 1991). In the Philippines, the *payao* continues to be a significant accessory to fishing, as is much discussed by all stakeholders involved, from fishers to politicians. In Thailand, the *sung* is used in the offshore pelagic fishery. Its usage is a much-studied topic among researchers, and the *sung* is popular among fishers as well.

INNOVATIONS TO ARTISANAL FADS IN SOUTHEAST ASIA: MODERNIZING AN OLD IDEA

It is interesting to note how small-scale fishers in the region have been using their ingenuity and local ecological knowledge to create innovative artificial reefs to assist their fishing operations in the Philippines, Thailand, and Indonesia. The three most interesting of these innovations are highlighted here: the lobster-fry, fan-like FADs called *pocong* in Lombok and Sumbawa, Indonesia (see Figure 6.2); the Juku Tech, the inexpensive but well thought-out innovation to the *rumpon* FAD in Sulawesi, Indonesia; and finally, the Tara-Bundu regulated *payao* in Timor-Leste.

The *pocong* is an ingenious artisanal FAD created to catch lobster fry. Although spiny lobsters were a known resource in Sumbawa and Lombok, the local fishers more often participated in the tuna fishery further offshore. However, this changed after 2010 when fishers opted for the less arduous option of exploiting the crustacean resources in their area. More specifically, they targeted the lobster fry found around these islands (Mongabay Indonesia 2018). To catch the lobster fry, a fan-like FAD is constructed of used cement bags cut into small pieces and folded into a disc-shaped,

FIGURE 6.2 Juvenile lobster FAD used in Lombok and Sumbawa, Indonesia. (Photo credit: Chang Fee Ming.)

fan-like unit. These smaller FADs are then are attached together along a string suspended from a small raft and left overnight in the calm coastal lagoons that serve as a natural nursery for the lobsters (*Panulirus* spp.). Fishers harvest the fry the next morning by shaking the fans that served as a shelter for the fry. These fry are sold to aquaculture farms for a lucrative amount. This novel, local innovation has been a boon to local fishers in these more remote parts of the Indonesian archipelago. However, since 2016, selling fry harvested from these modified fish aggregating devices is illegal: Indonesian fishers are no longer allowed to market any crustacean weighing less than 200 g outside Indonesia. Indeed, the Indonesian government has made it clear that Indonesian fishers must sell their harvest only in Indonesia. This has caused much contestation among fishers, many of whom continue to operate their lobster fry export business as usual. It is worth recalling the fallacy of the well-meaning assumption that fishers will willingly shift to more lucrative professions when presented with obstacles, as demonstrated by Pollnac et al. (2001). One fisher from South Lombok explained that he and his friends had tried to farm the fry they caught. However, he lamented that they were unable to grow their fry to the permissible weight. This is unlike Vietnamese aquaculture farmers who can successfully grow fry, without prior training, in a shorter time.

The Juku Tech FAD was developed in 2016 by a local, non-governmental organization called Sahabat Pulau (Friends of the Islands). This innovative FAD enabled the creation of a better-designed *rumpon* using an inexpensive, battery-operated telemetry system. Currently, this is the most frequently deployed FAD in Sulawesi. Moreover, the upgraded *rumpon* provides information to fishers connected to the internet. Such information is valuable to local fishers, as it will result in fuel savings.

The introduction of these innovations could cause concern among some resource managers as it may reignite the unresolved attraction versus production debate (IPNLF 2016, Yusfiyandayani et al. 2015). Even if the lobster fry caught using the *pocong* innovation were not exported directly and were transferred instead to local aquaculture farms, the development of this wild-sourced lobster aquaculture industry may still be damaging to the lobster population if uncontrolled. As for the Juku Tech *rumpon*, it is unknown if the fishers linked to the innovation are taking advantage of the situation by harvesting even more fish, now that they know exactly which *rumpon* to target. While these are valid concerns, they should not be considered under the pretext of ignorance that will lead fishers into a socio-ecological trap. Fishers have much to offer fisheries researchers and managers, owing to their traditional knowledge of the environment that has long guaranteed their fishery-dependent survival (Johannes et al. 2000). This knowledge has enabled fishers to adapt to challenges they face. However, it is also important to note that these valuable observations on the marine ecosystem were accumulated by these fishers over hundreds of years. Therefore, the adaptive behavior of local fishers is based on their traditional ecological knowledge, and is historically applicable to a relatively undisturbed environment. Hence, the application of traditional ecological knowledge to resource management in a relatively pristine marine ecosystem should be encouraged. For instance, in Timor-Leste where a local custom called *tara bandu* is used to manage the newly created deep-sea *payao*-type of FAD. The *tara bandu*-regulated *payao* is known to benefit the livelihoods of local fishers (Beverly et al 2012). Through this innovative hybridization between an old practice with a new resource, the coastal communities have been assisted in controlling the access of each fisher to the higher value of the newly explored tuna fishery. This has reduced user conflicts, and can serve as an example for the future of other community-based fishery management practices similar to the artificial reef community-based fishery management concept (see Mohd Jani et al. 2018).

It must be recognized, however, that some changes may have tragic consequences. This may well be the case when traditional ecological knowledge-based innovations are practiced in an environment in which the ecosystem regime has shifted (Mollman et al 2015). In such a scenario, the actions of fishers based on their traditional ecological knowledge generated under previous environmental conditions may no longer be valid. Evidently, the subtle changes of the vital elements that compromise ecosystem resilience are not easily noticeable by traditional resource users. Due to these missing "data," the fishers' determinations based on traditional ecological knowledge can be unreliable, and these fishers may unintentionally cause overfishing of some resources. But, as stressed by Boonstra and Hanh (2015), these socio-ecological errors should not be viewed as mismatches, but as part of the information verification processes. Therefore, if these data had been available to all stakeholders (including fisheries researchers and resource managers), perhaps recalculations could have been made and local fishing practices adjusted to ensure resource sustainability. In short, a more inclusive fishery involving a broad range of stakeholders may be the solution to this conundrum.

CONCLUSION

The extension of this review to include web-based information shared by the public and located via open search engines such as Google proved beneficial in filling the knowledge gap about artisanal FADs found in previously remote regions of Southeast Asia. This has allowed the review to be more inclusive and, thus, able to present a more realistic perspective of artisanal FADs in the region. From the Indian to the Pacific Oceans that frame the expanse of the Southeast Asian marine region, the application of artisanal FADs has been recorded. Called by different names, these structures share a strong commonality: built using inexpensive, locally available materials and serving as FADs for small-scale fishers. In many countries, they continue to be used despite the introduction of more modern artificial reef designs constructed of fewer natural materials since the 1970s. However, little attention has been given to their significance in local fisheries, despite their utility to many small-scale fishers who usually have fewer available financial resources to develop modern types of artificial reefs. Importantly, these artisanal FADs are more ecologically friendly to the marine environment due to the natural, organic materials of which they are composed. It also should be noted, however, that some of these unmonitored fish aggregating devices, particularly when deployed at an industrial scale, can cause concern among resource managers. But if the traditional, fundamental knowledge of the fishers upon which these FADs are based is recognized by fisheries researchers and managers, these devices can be merged with the available scientific knowledge to produce solutions that are not only likely to be successful in attaining management goals but will also be popular among resource users.

ACKNOWLEDGMENTS

The author thanks the organizers of ICFE 2019 for the invitation to present this paper at the conference. The research and administrative support given by the management of Universiti Malaysia Terengganu during the study is duly appreciated.

REFERENCES

Ali, A., Hassan, R., Bidin, R., and Theparoonrat, Y. 2011. Enhancing management of fishery resources through intensified efforts in habitat conservation and rehabilitation. *Fish for the People* 9(2):10–20.

Aprieto, V.L. 1991. Payao tuna aggregating device in the Philippines. In: *Symposium on Artificial Reefs and Fish Aggregating Devices as Tools for the Management and Enhancement of Marine Fishery Resources*, Columbo, 14–17 May 1990, ed. V.L.C. Pietersz, 1–15, Regional Office for Asia and the Pacific, Bangkok, Indo-Pacific Fish. Comm., FAO-UN 1991/11.

Beales, R. 1982. Investigations into fisheries resources in Brunei. *Monograph of the Brunei Museum Journal* 5:1–204.

Bergstrom, M. 1983. Review of experiences with and present knowledge about fish aggregating devices. *Bay of Bengal Programme*, BOBP/WP/23, Madras, India.

Beverly, S., Griffiths, D., and Lee, R. 2012. *Anchored Fish Aggregating Devices for Artisanal Fisheries in South and Southeast Asia: Benefits and Risks*. FAO Regional Office for Asia and the Pacific, Bangkok, Thailand, RAP Publication 2012/20, 65p.

Boonstra, W.J., and Hanh, T.T.H. 2015. Adaptation to climate change as social–ecological trap: A case study of fishing and aquaculture in the Tam Giang Lagoon, Vietnam. *Environment, Development and Sustainability* 17(6):1527–1544.

Brown, C.C. 1935. Trengganu Malay. *Journal of the Malayan Branch of the Royal Asiatic Society* 13(123): 1–111.

Brzeski, V.J., and Newkirk, G.F. 2000. *Lessons from the Lagoon: Research Towards Community Based Coastal Resources Management in Tam Gian Lagoon, Vietnam*, Dalhousie University, Coastal Resources Research Network, Halifax, NS.

Chou, L.M. 1997. Artificial reefs of Southeast Asia-do they enhance or degrade the marine environment? *Environmental Monitoring and Assessment* 44(1–3):45–52.

Delmendo, M.N. 1990. A review of artificial reefs development and use of fish aggregating devices (FADs) in the ASEAN region. In: *Symposium on Artificial Reefs and Fish Aggregating Devices as Tools for the Management and Enhancement of Marine Fishery Resources*, ed. V.L.C. Pietersz, 14–17.

Dempster, T., and Taquet, M. 2004. Fish aggregation device (FAD) research: Gaps in current knowledge and future directions for ecological studies. *Reviews in Fish Biology and Fisheries* 14(1):21–42.

Ebbers, T. 2003. Reconciling fishing and environmental protection: Resources enhancement strategies for the conservation and management of fisheries. *Fish for the People* 1(3):17–26.

Firth, R. 1975. *Malay fishermen: Their peasant economy*. Norton Library, New York.

Fishery and Aquaculture Country Profiles. Myanmar. 2006. Country Profile Fact Sheets. In: *FAO Fisheries and Aquaculture Department* [Online], Rome, Updated 2006. http://www.fao.org/fishery/ (accessed 4 January 2020).

Fishing Technology Equipments. Fish Aggregating Device (FAD). Technology Fact Sheets. Text by J. Prado. In: *FAO Fisheries and Aquaculture Department* [Online], Rome, Updated 27 May 2005. http://www. fao.org/fishery/equipment/fad/ (accessed 4 July 2019).

Ho, S.C. 1999. The brush park (Samrah) fishery at the mouth of the Great lake in Kampong Chhnang Province, Cambodia. In: *Present Status of Cambodia's Freshwater Capture Fisheries and Management Implications. Nine Presentations Given at the Annual Meeting of the Department of Fisheries of the Ministry of Agriculture, Forestry and Fisheries*, 19-21 January 1999, 79–89.

IPNLF. 2016. *Fish Aggregating Device Management*. IPNLF Issue Brief.

Johannes, R.E., Freeman, M.M., and Hamilton, R.J. 2000. Ignore fishers' knowledge and miss the boat. *Fish and Fisheries* 1(3):257–271.

Mohd Jani, J. 2015. Artificial reefs in Setiu: At A crossroad between tradition and modernity. In: *Setiu Wetlands: Species, Ecosystems and Livelihoods*, eds. F. Mohamad, J.M. Salim, J.M. Jani, and R. Shahrudin, Kuala Terengganu: Penerbit Universiti Malaysia Terengganu, 161–175.

Mohd Jani, J., Olson, E., and Patenaude, G. 2018. Re-exploring the application of artificial reefs for community-based fishery management in Malaysia. *American Fisheries Society Symposium* 86:235–249.

Möllmann, C., Folke, C., Edwards, M., and Conversi, A. 2015. Marine regime shifts around the globe: Theory, drivers and impacts. *Philosophical Transactions in Royal Society of London Series B Biological Science* 5:370.

Natsir, M., Widodo, A.A., Wudianto, W., and Agnarsson, S. 2018. Technical efficiency of fish aggregating devices associated with tuna fishery in Kendari fishing port–Indonesia. *Indonesian Fisheries Research Journal* 23(2):97–105.

Noranarttragoon, Sinanan, P., Boonjohn, N., Khemakorn, P., and Yakupitiyage, A. 2013. The FAD fishery in the Gulf of Thailand: Time for management measures. *Aquatic Living Resources* 26(1):85–96.

Parry, M.L. 1954. The fishing methods of Kelantan and Terengganu. *Journal of the Malayan Branch of the Royal Asiatic Society* XXXVII(II):77–144.

Pollnac, R.B., Pomeroy, R.S., and Harkes, I.H. 2001. Fishery policy and job satisfaction in three Southeast Asian fisheries. *Ocean and Coastal Management* 44(7–8):531–544.

Siriraksophon, S., and Sayan, S. 2016. Overview of current status and trends of fisheries and issues on resources enhancement in Southeast Asia. In: *Consolidating the Strategies for Fishery Resources Enhancement in Southeast Asia. Proceedings of the Symposium on Strategy for Fisheries Resources Enhancement in the Southeast Asian Region*, Pattaya, Thailand, 27–30 July 2015, Training Department, Southeast Asian Fisheries Development Center, Thailand, 26–34.

Srean, P. 2018. Factors influencing marine and coastal area situation in Cambodia. *Asian Journal of Agricultural and Environmental Safety* 1:12–16.

Supongpan, S. 2006. *Artificial Reefs in Thailand*. Southeast Asian Fisheries Development Center, Training Department. TD/TRB/74; 23 p.

Vinh, C.T. 2001. Assessment of relative abundance of fishes caught by gillnet in Vietnamese waters. In: *Proceedings of the Fourth Technical Seminar on Marine Fishery Resources Survey in the South China Sea, AREA IV: Vietnamese Waters, 18–20 September 2000.* Bangkok, Thailand, Secretariat, Southeast Asian Fisheries Development Center, 10–28.

Wiadnya, D.G.R., Damora, A., Tamanyira, M.M., Nugroho, D., and Darmawan, A. 2018. Performance of rumpon-based tuna fishery in the Fishing Port of Sendangbiru, Malang, Indonesia. *IOP Conference Series: Earth and Environmental Science* 139(1):12–19.

Yahaya, J. 1993, November. Fish aggregating devices (FADs) and community-based fisheries management in Malaysia. In: *Symposium on Socio-economic Issues in Coastal Fisheries Management*, 315–326.

Yusfiandayani, R. 2013. Fish aggregating devices in Indonesia: Past and present status on sustainable capture fisheries. *Galaxea. Galaxea, Journal of Coral Reef Studies* 15(Supplement):260–268.

Yusfiandayani, R., Baskoro, M.S., and Monintja, D. 2015. Impact of fish aggregating device on sustainable capture fisheries. *KnE Life Sciences* 224–237.

7 Design and Creation of Fishing Grounds in Japan with Artificial Reefs

Shinya Otake

CONTENTS

ABSTRACT

The earliest reports of artificial reefs in Japan occurred in the 1600s. As reported in an early document, Japanese artificial reefs were used to enhance fishing. For this reason, the early artificial reefs in Japan can be said to have been used as auxiliary fishing gear. This is closely related to the fact that Japanese people have long viewed the sea as an essential site for their sustenance. History indicates that using the sea as a place of leisure began only after 1945. In Japan, marine leisure activities have been around since 1960. These activities are broad and include surfing and yachting, largely derived from overseas, especially the United States. In Japan, the main national industry based on the sea is fisheries, followed by transportation and some natural resource development. Thus, the enhancement of fishing grounds through the deployment of artificial reefs has been actively conducted in Japan. In promotion of this activity, the compilation and dissemination of coastal fishing ground facility design guidelines and artificial reef fishing ground development plan guidelines have contributed greatly. In addition, there is a training program for engineers to promote adherence to these guidelines. Artificial reef deployments are supported by the excellent technical staff of the Japan Fisheries Agency, which also conducts evaluations of the performance of reefs for the Japanese fisheries administration. In terms of engineering, it is necessary that the objectives for the reef deployment be quantifiable. This is especially important as, legally, it must be determined if the objectives (in terms of the creation of a fishing ground) have been met. The Basic Law for Fisheries provides important oversight for the fishing industry. Moreover, the Japan Fisheries Agency specifies that each artificial reef deployment must be guided by a basic plan. It is necessary to have a basic objective when the creation of a fishing ground is proposed, and it is always important to determine what actual tasks need to be accomplished to achieve the objectives for creating the fishing ground. This chapter introduces design guidelines based on previous efforts, shows the basic concept of fishing ground construction, and indicates the direction to be taken in the future. In the future, economic engineering ideas derived from the management perspective of fishermen will help determine the usefulness of this technology in fishery management.

INTRODUCTION

In Japan, many and various types of artificial reefs have been deployed. Historically, fishermen's expectations for artificial reefs have been high, and there are also many requests from them to install reefs for the purpose of enhancing fishing grounds. In many cases, artificial reefs are expected to attract fish to a fishing area with the goal of catching the aggregated fish. It has become clear that artificial reefs are also effective in resource management, as they help protect or enhance the food chain and serve to limit fishing when they are deployed to inhibit fishers from some activities (e.g., trawling). For this reason, in Japan, artificial reefs are not only used as a type of fishing gear, but are sometimes used to provided protection to natural reefs and serve as an "off limits" no-fishing zone as a kind of marine protected area.

Based on the documented experience of deploying artificial reefs off Japan, fishing ground projects using artificial reefs were established more than 30 years ago. However, these projects were incomplete, as they did not include the maintenance management techniques that are being actively proposed today. In addition, the commercial benefit of an artificial reef had been determined based simply on the ratio of the commercial catch value to its construction cost. It is important to note that the technical development costs, that can often match the construction costs, were not included in these analyses. This refers to the inexactness of technical development costs, and as a result, a reasonable cost for maintenance could not be determined. In this chapter, while demonstrating the basic concept of applying artificial reefs for Japanese fishing grounds construction, a sound management strategy is proposed by reviewing the management of aquaculture at a small scale from an engineering perspective. From this examination, I suggest the appropriateness of fishery management using artificial reefs.

ARTIFICIAL REEFS' BUILDING HISTORY IN JAPAN

In the 1600s, the first evidence of artificial reefs to enhance fisheries in Japan was recorded (Kakimoto 2004). This record indicated that stone was added to a fishing ground and included evidence of a proposal seeking permission to sink an old ship as an artificial reef on these fishing grounds.

In the 1950s, artificial reefs were composed chiefly of concrete. In the 1970s, development of a large fishing ground using artificial reefs was initiated. The reason for the slow adoption of artificial reefs in fisheries management was because of enforcement of the law on the coastal fisheries in Japan. The most notable law is the Coastal Fisheries Grounds Enhancement and Development Program Act. In this Act, various technologies regarding artificial reefs were proposed. For example, large artificial reefs made of steel, upwelling structures, and improved seaweed beds in shallow seas were proposed. These deployments were made according to the needs of fishermen but based on the rules of the Act.

The rules, called the Orange Book (as the rules were compiled in an orange book!), served as the design guidelines for artificial reefs in the coastal fishing grounds off Japan (Zenshinkyo 1992). Additionally, the design guidelines for the creation of fishing grounds using artificial reefs (Zenshinkyo 2000) was named the Blue Book (the cover was blue). In these rules, the following required standards were proposed:

1. Safety: required safety for all materials, including concrete and steel. Basically, a dissolution test is performed to confirm that there is no toxicity.
2. Durability: the reef is required to function for at least 30 years.

To meet these standards, new methods to model waves and tidal currents were needed. Additionally, further designs unique to underwater structures were necessary, including design technology unique to artificial reefs. Special consideration was needed toward overcoming the impact to the artificial reefs during installation. Moreover, the designs needed to incorporate information relative to the nature of the target fish when designing artificial reefs (Nakamura 1985):

3. Reef design: artificial reef modules/units designed to enhance fish attraction, and life history attributes relative to the target fish species.
4. Placement plan: important for creating fishing grounds using artificial reefs.

Of these standards, Standards 1 and 2 can be accomplished by applying conventional engineering techniques. However, Standards 3 and 4 require statistical processing that depends on the nature of the fish. Here, Japanese scientists have conducted numerous studies to supply information regarding species-specific needs of fish on artificial reefs.

The cumulative total of fishing grounds created using artificial reefs up to 2001 in the Sea of Japan is shown in Figure 7.1. The vertical axis shows the volume of the fishing grounds in cubic meters and the horizontal axis shows the year. The graph indicates annual increases in the amount of reef material, but the overall total size of the reefs was only 27 m in height and 1 km in diameter during the first 15 years of the program.

In Japan, the creation of fishing grounds as a public work project stagnated despite local demand. Meanwhile, the demand for seaweed development projects and shallow sea fishing grounds increased. The reason for this is the change in age profiles among the fishermen. This is an important issue, as aging fishermen tend to avoid distant or dangerous fishing grounds.

The basic steps for creating fishing grounds are presented here. The first actions to take are to determine the target sea area and target organisms. This is done administratively with adjustments made in consideration of the specific fishery. The first two actions include:

A) site selection, and B) target marine organism selection.

The next step includes secondary action steps taken to include considerations of the environment:

C) classification of environmental factors, and D) removal or reduction of limiting factors to create an optimal environment for target species.

Nakamura (2015) introduced an analytical method called environmental resistance analysis. Figure 7.2 (a–d) shows the result of investigating the relationship between environmental conditions

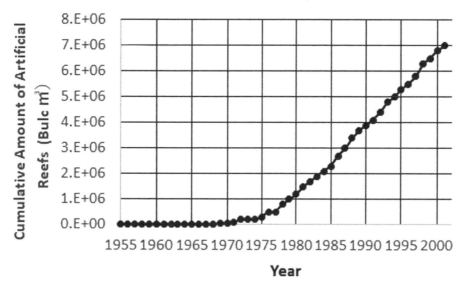

FIGURE 7.1 Cumulative number of artificial reefs along the Sea of Japan, off Japan. The vertical axis indicates the cumulative number of artificial reefs installed by bulk volume (cubic meters). The horizontal axis indicates the year.

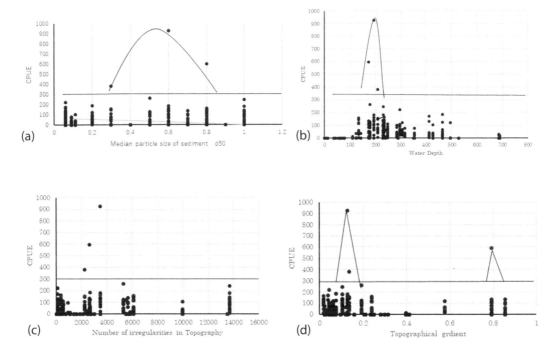

FIGURE 7.2 Relationships between CPUE of snow crab and various environmental conditions (filled circles). The curves show the preferred conditions for snow crab when the CPUE exceeded 300 crabs/operation: (a) Relationship between CPUE and median particle size of sediment (diameter in mm), (b) Relationship between CPUE and water depth in meters, (c) Relationship between CPUE and terrain roughness as indicated by the number of surface irregularities, (d) Relationship between CPUE and topographical gradient.

and the catch of snow crab taken during a survey in Wakasa Bay by using Nakamura's (2015) method of analysis (Otake, Kamide, and Kohno 2019).

The environmental conditions used here are seabed composition, such as bottom sediment, water depth, topographic gradient, and number of fishing reefs. As a result of this analysis, conditions for snow crab were favorable within a certain range shown within the curve in Figure 7.2. A multiple linear regression analysis showed that there was a positive correlation between the number of artificial reefs and the catch:

$$\text{CPUE} = -12.9 \times \text{D} + 2.11 \times \text{H} + 4.7 \times \varphi + 10.7 \times \text{N} + 27 \tag{7.1}$$

where D, H, φ, and N are shown in Table 7.1

Subsequent steps in the analysis involve the development and selection of technical methods and predictions of fish catch. The third group of action steps includes technology development to

TABLE 7.1

Environment Factors Examined Here and Their Definitions

Environmental factor	Definition
Median particle size of sediment (mm): 50	Average particle size in fishing area
Average depth(m): H	Average water depth in fishing area
Topographic gradient:φ	Maximum topographic gradient in fishing area
Number of irregularities in topography : N	3 m height difference within 200 m horizontal distance : for the total number of artificial reefs

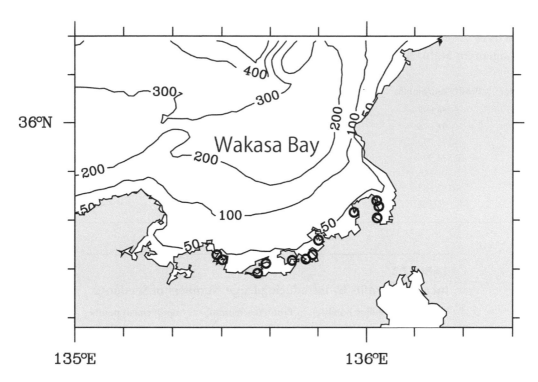

FIGURE 7.3 The location of the aquaculture farm in Wakasa Bay, the Sea of Japan, off Japan. The open circles indicate the location of the aquaculture farms.

preserve or create a favorable environment. The previous figure shows that artificial reefs are effective in increasing the fish population (either by attraction or production). Consequently, artificial reef installation is one of the technologies considered:

E) Development/choice of technical methods for assessment. For example, the availability of artificial reefs and wave control facilities.

A fourth group of action steps is the prediction of fish catch. This technology is in need of a new development of ecological model. Additionally, this technology needs to incorporate costs for maintaining target species. Consequently, we need to be able to calculate Life Cycle Costs:

F) Prediction of Fish Catch, and G) LCC (Life Cycle Cost) calculation of Technical Method

Development of a business model is part of the fifth group of action steps. The aquaculture industry in Obama, Fukui is composed of a group of small fisheries in Japan. The management strategy that maximizes profit by using linear programming for the aquaculture industry in the Wakasa Bay is considered here. The objectives are: 1) finding the number of seedlings that will maximize profits; and 2) determining the breeding conditions that maximize profits (Figure 7.3).

Figure 7.3 shows Wakasa Bay and the location of aquaculture farms in Wakasa Bay.

The aquaculture farmer's management status was calculated using linear programming (Otake and Hamza 2018). Table 7.2 indicates the calculated management status of seven farmers, who farm mainly tiger pufferfish, *Takifugu rubripes*. As a result, it was determined that five farmers were operating at a loss in spite of the fact that the growth rate and survival rate of their product are the same. Interestingly, when a farmer increased his purchase of seedlings, the management status of all farmers improved (Table 7.3). These results show that improvements in technology which lead to improvements in survival rate and growth rate of the target species, or increases in environmental capacity may not increase profitability. So this tells us that it is important to know if fishery improvements such as the installation of artificial reefs are useful for management.

TABLE 7.2

Management Status by Farmers of Tiger Pufferfish in Wakasa Bay, Fukui

Farmer	Profit (Yen/month)	Number of seedlings purchased	Environmental capacity	Growth rate	Survival rate
A	1,356,698	50,776	15,000	0.045	0.972
B	−94,676	0	5,250	0.061	0.980
C	−171,070	0	7,500	0.061	0.984
D	13,314,996	165,205	63,000	0.061	0.984
E	−101,816	0	3,000	0.045	0.972
F	−132,688	0	1,200	0.045	0.972
G	-83358	0	4050	0.045	0.979

TABLE 7.3

Increased Profits by Introducing Large Numbers of Seedlings

Farmer	Seedling number	Profit(Yen/month)	Improvement points
A	341,617	10,378,836	Same
B	122,468	3,505,421	Increased
C	956,756	52,108,736	Increased
D	1,116,392	108,436,650	Same
E	427,021	12,995,805	Increased
F	854,043	26,103,073	Increased
G	189,812	5,157,879	Increased

CONCLUSIONS

There are four main points that readers should understand as a result of this presentation:

1. Technological development is examined as objectively as possible.
2. Japanese technological development is aimed at utilizing the natural environment, and it is paramount that these efforts do not destroy the natural environment.
3. Much more can still be learned from nature.
4. Economic effects should not be neglected.

In Japan, there are have been many fishing grounds created by the deployment of artificial reefs, but these created areas do not dominate the catch. The reason for this is that types of fisheries and numbers of fish reefs are limited. For example, the purse seine and trawl fishery catches are large in Japan, while the catches are limited for the smaller fisheries such as line and gill net fishing that are often used on artificial reefs. Concomitantly, there are few fishers who participate in these smaller fisheries.

In contrast, the above-mentioned artificial reefs are benthic (i.e., they are deployed on the sea-floor). In Japan, the creation of fishing grounds by artificial reefs called floating artificial reefs or FADs (Fish Aggregating Devices) have been developed and have been deployed either 700 to 800 m deep or 1,000 m deep. The construction cost for FADs is lower than that of benthic artificial reefs, but currently the area of deployment, such as the Kuroshio coast, is limited. The useful life of a FAD is as short as 10 years, and maintenance efforts are considerable.

As mentioned above, the construction of fishing grounds using Japanese artificial reefs is examined in this chapter. It is important to note that the use of artificial reefs using concrete and steel has been limited, even though the historical use of artificial reefs goes back centuries. Currently, the scientific research efforts with regard to artificial reefs are significant, but there is some opposition to expanding artificial reef deployments by older fishermen. We believe that the demand for artificial reefs will increase with a decline in number of the older fishermen in the fishery.

REFERENCES

Kakimoto, H. 2004. Artificial fish reef. *The Japanese Institute of Technology Fishing Port Grounds and Communities*, 62p, Tokyo, Japan, (in Japanese).

Kohno, N., S. Okubo, S. Otake, W. Fujii, T. Kaneko, T. Miyamukai, and K. Imao. 2018. Environmental factor analysis of snow crab associated with artificial reefs. In: *Marine Artificial Reef Research and Development: Integrating Fisheries Management Objectives*, ed. S. A. Bortone. American Fisheries Society, Symposium 86, Bethesda, MD, 69–80.

Nakamura, M. 1985. Evolution of artificial fishing reef concepts in Japan. *Bulletin of Marine Science* 37:271–278.

Nakamura, M., and S. Otake. 2015. Enhancement of marine habitats. *International Compendium of Coastal Engineering*, 375–396, World Scientific Publishing Co. Pte. Ltd, Singapore.

Otake, S., and T. Hazama. 2018. Management engineering study of puffer fish farmers in Wakasa Bay. *Proceedings of Annual Meeting of the Japanese Society of Fisheries Engineering*, 27–30 (in Japanese).

Otake, S., J. Kamide, and N. Kohno. 2019. Construction of catch model considering behavioral ecology of Snow Crab in Wakasa Bay. *Proceedings of Annual Meeting of the Japanese Society of Fisheries Engineering*, 5–8, JSFE, Tokyo, Japan, (in Japanese).

Zenshinkyo. 1992. Design guidelines for coastal fishing ground creation facilities. Zenkoku Engan Gyogyo Sinko Kaihatu Kkyokai Co. Ltd., Tokyo, Japan, (in Japanese).

Zenshinkyo. 2000. Design guidelines for fishing ground creation using artificial reefs. Zenkoku Engan Gyogyo Sinko Kaihatu Kyokai Co. Ltd., 187p., Tokyo, Japan, (in Japanese).

8 Using Standardized CPUE to Estimate the Effect of Artificial Reefs on Fish Abundance

Nariaki Inoue, Satoshi Ishimaru, Kengo Hashimato, Junji Kuwamoto, Takahito Masubuchi, and Minoru Kanaiwa

CONTENTS

ABSTRACT

Catch per unit effort (CPUE), calculated from data obtained from commercial fisheries, is widely used for resource assessment. However, the untransformed nominal CPUE data are affected by many factors (e.g., natural annual and seasonal fluctuations, differences in fishing areas and fishing gear). Consequently, using the nominal CPUE data does not allow an accurate evaluation of the effect that an artificial reef may have on fish stocks. Standardized CPUE, which is considered as a stock abundance index, is a fundamental component of fishery stock assessment. In this study, we apply the standardized CPUE method for removing the same factors' impact, except for that of the artificial reef, from the nominal CPUE to estimate the effect of artificial reefs on fish abundance for Japanese butterfish (*Hyperoglyphe japonica*).

INTRODUCTION

Artificial reefs are thought to improve fishing by promoting fish aggregation and functioning as both nursery areas and habitats for juveniles, thereby increasing fisheries stocks in terms of diversity and abundance (Bohnsack and Sutherland 1985). From 1976 to 2001 in Japan, a government subsidy program to enhance coastal fishing grounds was established that included incentives for artificial reef deployments. The Japanese government has spent at least 1.86 trillion JPY on the program for artificial reef construction solely to improve fin fish fisheries (Yamane 1989).

Although there are numerous studies on artificial reefs, few studies have investigated their benefit to fisheries through quantitative, experimental methods (Bohnsack and Sutherland 1985).

Catch per unit effort (CPUE), calculated from data obtained from commercial fisheries, is widely used for stocks assessment. Many studies on the influence that artificial reefs have on fisheries have used fishery-independent data to assess the magnitude of this influence, but few studies have made use of data from commercial fisheries (i.e., fishery-dependent data). The lack of inclusion of commercial fisheries and concomitant CPUE information in this assessment is due to numerous factors. These factors include: natural annual and seasonal fluctuations in the stocks, differences in fishing areas, and differences in capture efficiency between fishing gear. Consequently, using the nominal (i.e., uncalibrated or untransformed) CPUE data does not allow an accurate evaluation of the effect that an artificial reef may have on fisheries. Here, the focus was on fluctuations in stocks using standardized CPUE information derived from generalized linear models (GLMs) (Okamura et al. 2018). This study considers annual trends that provide a representative abundance index calculated by removing the influence of the varying factors that bias the nominal CPUE trend (Okamura et al. 2018) and, subsequently, influence our ability to obtain an understanding of relative fish abundance (as derived from a presumed relationship that fish abundance has with CPUE). Here standardized CPUE was used to more reliably estimate the effect of artificial reefs on the relative abundance of fishery stocks as determined in a previous stock assessment by Inoue et al. (2018).

CPUE STANDARDIZATION

If the catch (C) is proportional to the product of the stock (N) and the fishing effort (E), then:

$$C = q \cdot E \cdot N \tag{8.1}$$

Considering that CPUE is proportional to abundance N, and q is fishing efficiency, then:

$$CPUE = C/E = q \bullet N \tag{8.2}$$

Assuming the fishing efficiency (q) may change depending on the fishing area (q_a), season (q_s) gear (q_g) and others. The CPUE will also change under the constant N because of the influence of these changing variables, therefore:

$$CPUE = C/E = q \bullet q_a \bullet q_s \bullet q_g \bullet N \tag{8.3}$$

Consequently, it is necessary to remove the effect of q_a, q_s, and q_g on the nominal CPUE data to estimate a Standardized CPUE that has the effects of these variables removed. Moreover, this Standardized CPUE would be able to serve as an unbiased estimate of the relative abundance of a fishery. In general, fluctuations in the Standardized CPUE (sCPUE) better reflects the actual stock fluctuation than nominal CPUE. A general explanation for performing CPUE standardization is described by Hilborn and Walters (1992) and Quinn and Deriso (1999).

Generalized Linear Models (GLMs) are the most popular approach for achieving CPUE standardization, as it allows one to estimate the value of various factors (e.g., q_a, q_s, and q_g) that influence the nominal CPUE, and then remove the influence of those effects (Shono 2004).

GENERALIZED LINEAR MODELS

To standardize the CPUE, GLMs are determined for the factors influencing CPUE (e.g., year, season, fishing area, and presence or absence of an artificial reef) as explanatory, independent variables on the right side of the equation and the CPUE response as a dependent variable on the left side of the equation. Typically, stock assessment studies focus on annual fluctuations in the fish stock (e.g., effect of the year on CPUE); however, here we focused on the effect of the presence/absence of the artificial reef on relative fish abundance as presumed by examining a standardized CPUE. There are many published reports that explain the relationship between stock assessments using statistical models such as GLMs (e.g., Okamura and Ichinokawa 2016). Using the GLMs, the right side of the equation is described as a simple linear equation (linear predictor), and this estimated equation with a presumed normal distribution is referred to as the CPUE-LogNormal model:

$$\log\left(\text{CPUE}\right) = \beta_0 + \beta_1 \bullet \text{Year} + \beta_2 \bullet \text{Season} + \beta_3 \bullet \text{Area} + \beta_4 \bullet \text{AR} \qquad (8.4)$$

For determining the CPUE using a GLM model, the coefficients of each explanatory variable (β_{1-4}) were estimated. Coefficients β_1 and β_2 were regarded as yearly and seasonal trend indexes of the stock, β_3 was the natural fish distribution trend that was isolated from the artificial reef effect, and β_4 was regarded as the artificial reef effect, respectively. However, a CPUE-LogNormal model cannot be applied to zero-catch data (i.e., CPUE = 0) because the natural logarithm of zero-catch data is minus infinity. To compensate for this, three methods have often been applied (e.g., Shono 2008): (1) adding a small constant value (e.g., 0.1) to all CPUE values in the dataset; (2) using a GLM with a Poisson or negative binomial error structure (Reed 1996); or (3) using the delta-type, two-step model (Lo et al. 1992). With the first method, there is no setting criteria for adding constant value and the adding effect is unclear. Additionally, CPUE data was obtained by weight (Kg) base in this study, thereby not assuming for a Poisson error structure that needs countable data (e.g., individual number of fish catch). Accordingly, this study used the delta-type, two-step model to estimate the effects of artificial reefs on the abundance of *H. japonica*.

Japanese butterfish (*Hyperoglyphe japonica*) is a commercially important species in coastal northwest Pacific Ocean fisheries and is chiefly caught by small vessels using pole-and-line fishing in the waters off the Iki Islands, Nagasaki Prefecture, Japan (Figure 3.1). This study used the delta-type, two-step model to estimate the effects of artificial reefs on the density of *H. japonica* in the waters around these islands (Figure 8.1).

ESTIMATION OF THE ARTIFICIAL REEF EFFECT

DATASET

We used two years of fishing records (2014 and 2015) collected in Nagasaki Prefecture from four small ships (4.8–6.6 tons) operated around the Iki Islands. The ships fished on one or more sites associated with an artificial reef (distance between each fishing site and the nearest artificial reef was <200 m) or on sites associated with a natural reef (>200 m) during a cruise completed in one day. The data recorded included information on the amount of fishing time, fishing site (Lat, Lon), depth, and reef type (near either an artificial reef or a natural reef). However, the weight of each fish species caught was recorded only per cruise for Kg base, and therefore the unit of analysis was fish weight per cruise. Estimation of the influence of the presence of an artificial reef as well as the fishing time per cruise (T-time), the fishing site (i.e., mean Lat, Lon per site), the mean fishing depth (Depth), and the ratio of the fishing time on the artificial reef (A) in relation to the total fishing time per cruise (i.e., A/T-time) on the CPUE as determined by the GLMs that were calculated for each cruise.

UTILIZATION OF ARTIFICIAL REEFS AROUND THE IKI ISLANDS

During 2014 and 2015, the four small fishing ships completed a total of 520 and 567 cruises, respectively, around the Iki Islands, and included 1,940 fishing sites in 2014 (1,033 artificial reef associated sites; 907 natural reef associated sites) and 2,364 fishing sites in 2015 (979 artificial reef associated sites; 1,385 natural reef associated sites). Approximately 47% of the fishing was conducted in artificial reef associated areas; much of the fishing effort occurred in the western waters around the islands (Figure 8.1). In contrast, a large proportion of the fishing occurred in natural reef-associated

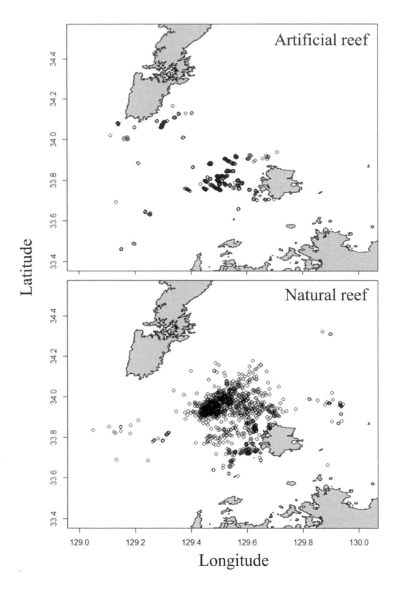

FIGURE 8.1 Maps of the fishing sites for four vessels in the waters off the Iki Islands, Nagasaki, Japan during 2014 and 2015. Open circles of upper and lower figures indicate the fishing sites on artificial and natural reef area, respectively (from Inoue et al. 2018).

areas in northwestern waters, as this region is considered to have good fishing grounds on this ocean bank area (the Shichiriga-sone Bank).

The total catch in 2014 and 2015 was 15,496 kg and 18,537 kg, respectively, and the main species caught were *H. japonica* (comprising 27.8% of the total catch by weight during the two-year fishing period), *Seriola quinqueradiata* (28.2%), *Thunnus orientalis* (11.8%), and *Pagrus major* (9.7%). The dataset included calculated values for A/T-time only for cruises that fished sites with artificial reefs (47.1% of all cruises) and zero for those cruises that fished only natural reefs (36.2%), selected from all data and compared to the species composition of catch (Figure 8.2). When catches were obtained from cruises operating over natural reef sites, *S. quinqueradiata* (40.5%) and *T. orientalis* (11.8% of the catch by weight) were dominant. When fishing was conducted in artificial reef-associated areas, *H. japonica* comprised 70% of the catch by weight, and this was 3.56 times more than their weight in the total catch obtained from cruises operated in the natural reef areas. This indicates that fishing areas near artificial reefs are important fishing grounds for *H. japonica*.

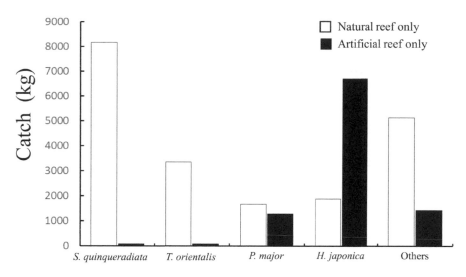

FIGURE 8.2 Fish species composition caught by fishing over two reef types: fishing on artificial reef areas (black bars) and on natural reef areas (white bars) (from Inoue et al. 2018).

DELTA-TYPE, TWO-STEP MODEL

In this study, the delta-type, two-step model (Lo et al. 1992) was used calculate the sCPUE to serve as an index of *H. japonica* relative abundance in association with artificial reefs. In Step 1 of this model, the positive catch probability (P) was estimated by a logistic model. In Step 2, the positive catch cruise data were selected, and a log-based, positive CPUE was estimated by using the CPUE-LogNormal model.

Step 1:

Positive catch probability (P) was estimated using GLM as the response variable with Year, Month, Ship, Lat, Lon, Depth, log T-time, and A/T-time as explanatory (independent) variables for the initial model. The error structure assumed a binomial distribution and used logit for the link function (left side of equation) for the initial model.

$$\log(P/1-P)) = \beta_0 + \beta_1 \cdot \text{Year} + \beta_2 \cdot \text{Month} + \beta_3 \cdot \text{Ship} + \beta_4 \cdot \text{Lon}$$
$$+ \beta_5 \cdot \text{Lat} + \beta_6 \cdot \text{Depth} + \beta_7 \cdot \log(\text{T-time}) + \beta_8 \cdot \text{A/T-time} \tag{8.5}$$

In the eight explanatory variables included in the initial model, an optimal combination of explanatory variables for the logistic model was selected by using some statistical indexes (e.g., Akaike's information criterion (AIC) and the corrected AIC (AICc), and Bayesian information criterion (BIC)). In the present study, the BIC was used to select an optimal model (Model 1). The explanatory, dependent variables selected for the Step 1 optimal model for *H. japonica* (Model 1) were: Ship, Lon, log (T-time), and A/T-time. This analysis indicated that the positive catch probability (P) of *H. japonica* was statistically associated with A/T-time (Table 8.1). Using Model, this outcome (Figure 8.3) was produced under the assumptions for the average operation during a cruise (i.e., substitute the two response variables Lon and log(T-time) for the average values; and change the values of A/T-time within the range 0 to 1) to estimate the effect of A/T-time on positive catch probability (P). When catches were obtained only in association with artificial reefs or natural reefs, the positive catch probability (P) for *H. japonica* was 0.78 and 0.35, respectively.

Step 2:

The positive-catch cruise data were selected (415 out of 1,087 cruises), and log-based positive CPUE (log (CPUE$_{posi}$)) was estimated for the same response variable as in Step 2 of the model. The error

TABLE 8.1

Likelihood Ratio Test by Chi-Square Test of the Optimal Generalized Liner Models (GLM) for Changes in Catch Probability and CPUE (>0) for Two Models

Species	Model	Exp. var.	Df	Resid. dev.	X^2 value
Hyperoglyphe japonica	Model 1	Null		1201.180	
		Ship	3	1047.490	96.111 ***
		Lon	3	950.670	53.064 ***
		log *(T-time)*	1	1032.560	27.876 ***
		A-time rate	1	853.580	97.091 ***
	Model 2	Null		293.900	
		A-time rate	1	282.090	11.193 **
		log *(T-time)*	1	287.770	7.625 ***

df, degree of freedom.
*$p <0.05$; **$p <0.001$; ***$p <0.0001$

structure of Step 2 of the model assumed a normal distribution and used it to identify the link function (right side of equation) with the initial model, and the optimal model (Model 2) selected by BIC:

$$\log\left(\text{CPUE}_{\text{posi}}\right) = \beta_0 + \beta_1 \cdot \text{Year} + \beta_2 \cdot \text{Month} + \beta_3 \cdot \text{Ship} + \beta_4 \cdot \text{Lon}$$
$$+ \beta_5 \cdot \text{Lat} + \beta_6 \cdot \text{Depth} + \beta_7 \cdot \log\left(\text{T-time}\right) + \beta_8 \cdot \text{A/T-time} \tag{8.6}$$

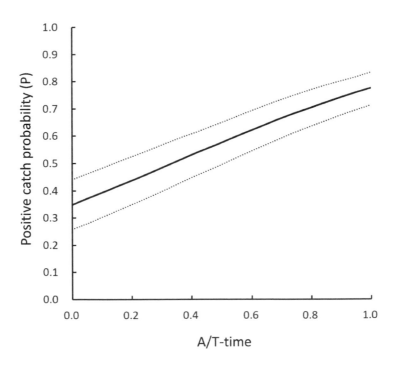

FIGURE 8.3 Estimated positive catch probability (P) curves for *Hyperoglyphe japonica* in relation to the ratio of the total operation time on the artificial reef to the T-time (i.e., A/T-time) during each cruise. The solid curve indicates the positive catch probability (P), and the upper and lower dotted curves indicate the 95% confidence interval.

The response variables selected for Step 2 in the optimal model for *H. japonica* (Model 2) were log (T-time) and A/T-time (Table 8.1). The results indicated that the positive CPUE ($CPUE_{posi}$) was associated with A/T-time. Using Model 2 (with the same assumptions in Model 1), the relationship of A/T-time to $CPUE_{posi}$ was estimated (Figure 8.4). The $CPUE_{posi}$ calculated for *H. japonica* obtained from cruises conducted only in either artifical reef or natural reef areas was 17.06 kg and 12.7 kg, respectively.

Finally, under the conditions of a typical cruise (i.e., created by removing the fluctuation effects of year, month, location, amount of fishing time, and depth) sCPUE was calculated by multiplying the two estimation values calculated from the Step 1 and 2 optimal models (Figure 8.5).

Discussion: Estimation of the Artificial Reef Effect

When the cruise was conducted only in areas with natural reefs, the sCPUE for *H. japonica* was 4.44 kg and that value increased with increasing A/T-time, and attained 13.24 kg when A/T-time = 1 (interpolated as if the operating was conducted only at artificial reef associated sites, Figure 8.4). In comparison to results from catches obtained from natural reef areas, we note about a three-fold increase in fish abundance from catches obtained for artificial reef associated areas. According to Moriwaki et al. (2005) and Yamasaki ct al. (2013), who studied fisheries and distribution patterns on fishing grounds in the waters off Kyoto and Shimane Prefectures, respectively, the *H. japonica* catch and untransformed CPUE at artificial reef-associated sites was significantly higher than the catch and CPUE at natural reef-associated areas, in concurrence with the results reported here. In conclusion, the presence of artificial reefs in fishing grounds off Japan had a positive effect on the CPUE of *H. japonica* in comparison to the CPUE for the species in areas lacking in artificial reefs.

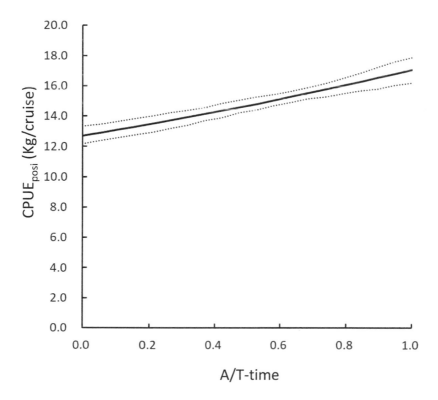

FIGURE 8.4 Estimated positive catch $CPUE_{posi}$ curves for *Hyperoglyphe japonica* in relation to the ratio of the total operation time on the artificial reef to the T-time (A/T-time) at each cruise. The solid curve shows the $CPUE_{posi}$, and the upper and lower dotted curves show a 95% confidence interval.

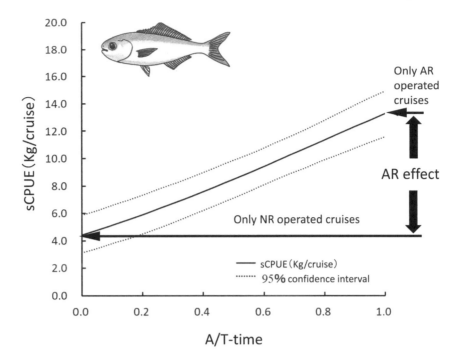

FIGURE 8.5 Estimated sCPUE curves for *Hyperoglyphe japonica* in relation to the ratio of the total time on the artificial reef to the T-time (A/T-time) during each cruise. The solid curve indicates the sCPUE, and the upper and lower dotted curves indicate a 95% confidence interval (modified from Inoue et al. 2018).

We speculate that there are two factors to explain the increase in fish abundance, each of which may occur independently or synchronously. First, an increase in fishing efficiency (q) may occur due to the behavioral aggregation response of fish to artificial reefs. Second, there may be an increase in the stock (N) (see Equation (8.3)) (Polovina 1989, Pratt 1989, Seaman et al. 1989). Although it is well-known that artificial reefs have an aggregation effect on some fish species (Polovina 1989), there are only a few studies that conclude that artificial reefs result in an increase in fishery stocks. According to Huntsman (1981) and Plovina (1989), artificial reefs were ineffective in increasing stocks at the fisheries level. It cannot be determined from our study whether or not the artificial reef resulted in an increase of *H. japonica* in the waters of Iki Islands. To further elucidate this situation, a detailed CPUE comparison from areas with artificial reefs and natural reefs is required (Carr and Hixon 1997). Furthermore, future investigations should include an analysis of the temporal changes in CPUE which include before and after deployments of artificial reefs (Matsumiya et al. 1991).

REFERENCES

Bohnsack, J. A., and D. L. Sutherland. 1985. Artificial reef research: a review with recommendations for future priorities. *Bulletin of Marine Science* 37:11–39.

Carr, M. H., and M. A. Hixon. 1997. Artificial reefs: the importance of comparisons with natural reefs. *Fisheries* 22(4):28–33.

Hilborn, R., and C. J. Walters. 1992. *Quantitative Fisheries Stock Assessment*. Chapman and Hall.

Huntsman, G. R. 1981. Ecological considerations influencing the management of reef fishes. In *Artificial Reefs: Conference Proceedings*, ed. D. Y. Aska, 167–175. Florida Sea Grant College Report 41, Gainesville, FL.

Inoue, N., Nambu, R., Kuwahara, H., Kuwamoto, J., Yokoyama, J., and M. Kanaiwa. 2018. The present situation of utilization and the effects of artificial reefs on the resource density of Japanese butterfish *Hyperoglyphe japonica* and red seabream *Pagrus major* in the waters of the Iki Islands, Nagasaki, Japan. *NIPPON SUISAN GAKKAISHI* 84:1010–1016.

Lo, N. C., Jacobson, L. D., and J. L. Squire. 1992. Indices of relative abundance from fish spotter data based on Delta-Lognormal models. *Canadian Journal of Fisheries and Aquatic Sciences* 49:2515–2526.

Matsumiya, Y., Oka, M., Hiramatsu, K., and K. Asano. 1991. A theoretical study on the calculation of artificial reef. *Bulletin of the Faculty of Bioresources Mie University* 5:73–77.

Moriwaki, S., Tameishi, T., Wakabayashi, H., Matumoto, H., Tamaka, N., and H. Saito. 2005. Changes in catch composition of the gathered fish to High-rise artificial reef off Hamada. *Report of Shimane Prefectural Fisheries Technology Centre* 12:1–6. (In Japanese)

Okamura, H., and M. Ichinokawa. 2016. Statistical modelling in fisheries science. *Proceedings of the Institute of Statistical Mathematics* 64:39–57. (In Japanese with English abstract)

Okamura, H., Morita, S. H., Funamoto, T., Ichinokawa, M., and S. Eguchid. 2018. Target-based catch-per-unit-effort standardization in multispecies fisheries. *Canadian Journal of Fisheries and Aquatic Sciences* 75:452–463.

Polovina, J. J. 1989. Artificial reefs: nothing more than benthic fish aggregators. *CalCOFI Reports* 37:37–37.

Pratt, J. R. 1989. Artificial habitats and ecosystem restoration: management for future. *Bulletin of Marine Science* 55:268–275.

Quinn, T. J. II, and R. B. Deriso. 1999. *Quantitative Fish Dynamics.* Oxford University Press, New York, NY.

Reed, W. J. 1996. Analyzing catch-effort data allowing for randomness in the catching process. *Canadian Journal of Fisheries and Aquatic Sciences* 43:174–186.

Seaman, W., Buckley R. M., and J. J. Polovina. 1989. Advances in knowledge and priorities for research technology and management related to artificial aquatic habitats. *Bulletin of Marine Science* 44:527–532.

Shono, H. 2004. A review of some statistical approaches used for CPUE standardization. *The Japanese Society of Fisheries Oceanography* 68:106–120. (In Japanese with English abstract)

Shono, H. 2008. Confidence interval estimation of CPUE year trend in delta-type two-step model. *Fisheries Science* 74:712–717.

Yamane, T. 1989. Status and future plans of artificial reef projects in Japan. *Bulletin of Marine Science* 44:1038–1040.

Yamasaki, A., Tsuji, S., and Y. Hamanaka. 2013. Estimates of the main fish caught by recreational fishing boats in coastal waters off Kyoto Prefecture. *Report of Kyoto Prefectural Agriculture, Forestry and Fisheries Technology Center* 35:25–46. (In Japanese)

9 Using Artificial Substrata to Recover from the Isoyake Condition of Seaweed Beds off Japan

*Osamu Hashimoto, Motobumi Manabe, Akira Watanuki,
Masaru Kawagoshi, Takeshi Hosozawa, Fumihisa
Okashige, Yasuyuki Gonda, Syouichi Ito, Takeshi
Tajima, Yousuke Fukui, Tomomi Terajima, Hirokazu
Nishimura, Tetsuya Shirokoshi, and Toru Aota*

CONTENTS

ABSTRACT

In many coastal areas in Japan, the rocky-shore seaweed denudation called "isoyake" has been a major problem. This chapter examines recovery methods for the re-establishment of seaweeds through an examination of the materials and activities necessary to promote the development of healthy seaweed beds. This chapter also discusses measures that facilitate continued maintenance for seaweed bed restoration. These measures generally include defenses against herbivores, planting, fertilization, and deploying substrates. Success in these efforts depends upon cooperation between construction efforts and refinement of fisheries regulations. Future success in this effort should include both structural and non-structural countermeasures.

INTRODUCTION

Recently, a large-scale disappearance of seaweed beds called "isoyake" has been observed in many coastal areas throughout Japan, with a concomitant impact on the Japanese fishing industry. "Iso" is the Japanese word for rock and "yake" means burnt field. The barren ground or rocky-shore denudation is called isoyake because the seaweed bed appears to have been burned by fire. In an isoyake seaweed bed, commercially important aquatic mollusks such as abalone and turban shell have been decreasing, which is having a problematic impact on many coastal fisheries. Although various causes have been postulated for the occurrence of isoyake, it has become clear that the effects of predation on seaweed from sea urchins and herbivorous fish is significant.

In 2007, the Japan Fisheries Agency published the *Isoyake Recovery Guidelines* (Japan Fisheries Agency 2007), which indicated that controlling the density of aquatic herbivorous animals is important in the maintenance of fisheries. These guidelines summarize methods for diagnosing marine conditions and what measures should be taken to remedy various situations. Since the publication of the guidelines, the need for recovering from the isoyake condition has been appreciated. In addition, since some of the elemental technologies listed in the guidelines have evolved, the guidelines were revised in 2015 (Fishery Agency 2015). The revised guidelines indicate that it is important to generate a phase shift in the balance between herbivore feeding pressure and seaweed growth to facilitate recovery from the isoyake condition. These guidelines focus on non-structural measures in the form of techniques for fishers to implement to restore the seaweed beds. Conversely, there are no detailed descriptions in the guidelines for structural measures to be implemented by local governments, such as the installation of artificial substrata for the formation of seaweed beds.

Therefore, our study group, organized by private companies, brought together the technologies and products developed and owned by each participating company and categorized the various methods that showed promise in promoting recovery from the isoyake condition of the seaweed beds. This chapter will examine these recovery methods through an examination of the materials and activities necessary to promote the development of healthy seaweed beds.

MATERIALS AND METHODS

In Japan, there are dozens of private companies involved in the creation of seaweed beds. The group enlisted here consisted of ten representative companies. By organizing the technologies developed and owned by each company, most of the physically applied countermeasure technologies known to date in Japan have been included for consideration in recovery efforts. In the isoyake recovery guidelines published by the Japan Fisheries Agency, the elemental technologies for recovery of seaweed beds were organized according to the factors that inhibit the formation of seaweed. Therefore, 20 countermeasure technologies developed by the companies enlisted here were examined and classified according to the manner in which they affect the formation of seaweed beds, as indicated in the *Isoyake Recovery Guidelines*. The introduction of these technologies, with examples of the recovering seaweed beds, were compiled in a booklet entitled "Measures Against Isoyake Using Artificial Substrata." This booklet suggests that good coordination between structural and non-structural measures is important for recovery from the isoyake condition.

RESULTS

Figure 9.1 indicates the physical countermeasure technologies employed according to the factors that inhibit the formation of seaweed beds. Each technology was broadly divided into measures against damage caused by herbivores (Reduction of Grazing) and techniques for increasing seaweed (Stock Enhancement). Some of the countermeasure technologies are elaborated below.

DEFENSE AGAINST HERBIVORES

Defense techniques enumerated here include protection against feeding damage by sea urchins and herbivorous fishes.

The shallower the water depth, the greater the influence of wave action, to the point where sea urchins can no longer consume the seaweed. As measures against feeding damage by sea urchins,

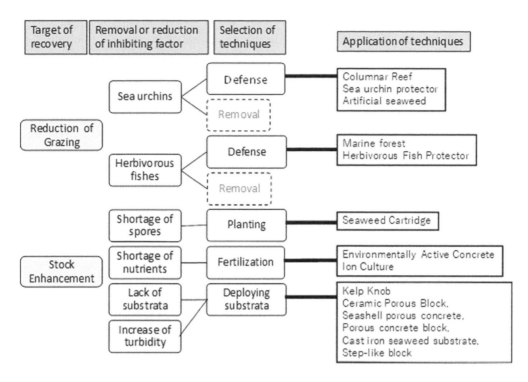

FIGURE 9.1 Organizational strategy for reducing the isoyake condition on seaweed beds.

concrete columnar artificial reefs (i.e., Columnar Reefs) were developed to increase wave action (i.e., increase water velocity) by reducing the water depth and, consequently, reduce feeding pressure by sea urchins (Figure 9.2). Columnar Reefs were designed as tall artificial reef modules to protect the seaweed. If the water depth at the top of the columns is reduced, most sea urchins cannot inhabit the top of the structure (owing to increased effects of wave action), allowing the seaweed to survive at the top of the columns. The deployment of these Columnar Reefs is simply the safest and most effective way to create sustainable seaweed beds. In sea areas where many herbivorous fish are distributed, attempts are currently being made to protect the columns with cages and resin spikes to reduce feeding pressure by herbivorous fishes.

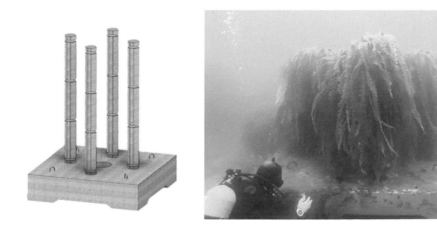

FIGURE 9.2 Columnar Reef that restricts sea urchins to lower portions of the water column.

In addition, needle mats were deployed to prevent sea urchins from entering areas where seaweeds are likely to grow. By orienting the needle mat downward on the top of the outside of exclusion fences (Figure 9.3), the sea urchins cannot easily enter, so the aquatic vegetation inside the fenced area is protected from predation. Presently, there are many examples of protecting corals from the damage caused by black longspine urchins (Figure 9.3). Recently, to protect the kelp beds, sea urchin trap experiments were conducted using exclusion fences with needle mats inside. The results of these experiments indicated the traps were effective in protecting the kelp from predation. Various future applications are expected to be developed as a result of the experiments and procedures indicated here.

As a structure for protecting seaweeds from herbivorous fish, caged blocks have long been modified for this purpose (Figure 9.4). It is expected that seaweed seeds will be supplied from seaweeds in the cage to the surrounding area. The probability of success is high for such a simple structure, but success cannot be achieved with this deployment alone. Other non-structure, active measures, such as maintenance by fishers, are required to assure future success when using these structures. To increase the effectiveness of caged blocks in promoting seaweed recovery, it is necessary to periodically clean the accumulated attached organisms from the cages and thoroughly remove herbivorous fish from the area.

FIGURE 9.3   Needle Mat to prevent sea urchins from entering a seaweed growth area.

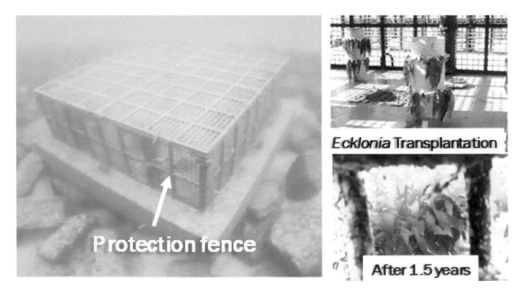

FIGURE 9.4   Block with fiber-reinforced plastic cage to protect against herbivorous fish.

FIGURE 9.5 Herbivorous Fish Protector protects growing points of seaweed from herbivorous fishes.

A spindle device mounted on blocks was developed to protect seaweeds from being preyed upon by herbivorous fish (Figure 9.5). By attaching several spine-like defenses (i.e., the Herbivorous Fish Protector) on a block, the seaweed growth points can be protected from herbivorous fish. Rather than protecting the whole seaweed, the Protector provides the minimum protection necessary for the plant parts (i.e., juvenile seaweed and seaweed growth points) to assure its survival. The Herbivorous Fish Protector is a less expensive structure that can facilitate seaweed growth more efficiently than caged blocks.

PLANTING

If the decline of seaweed extends over a wide area, there may be a shortage of mature "mother" algae to supply spores or embryos. In such areas, it is necessary to plant seaweed spores and embryos or transplant seaweed seedlings. With this method, spores or embryos are sprinkled on small plates and seaweed on the plates sprout after several months. These plates bearing sprouts are then attached to stable blocks which subsequently allow for effective seaweed growth. Recently, a removable substratum, Seaweed Cartridge (Figure 9.6), has been developed. Conventional plates (artificial substrata) normally cannot be removed once they are fixed to the block. On the other hand, the Seaweed Cartridge substratum can be removed and reused many times. If, for some reason, the seaweed on this removable substratum declines, the cartridge can be retrieved and reused. Furthermore, if the seaweed on the Seaweed Cartridge dies, the cartridge can easily be replaced with another Seaweed Cartridge on which healthy seaweed is growing. As a result, there is a high probability that the sustainable seaweed bed will increase in growth.

FIGURE 9.6 Seaweed Cartridge, a small reusable substratum.

FERTILIZATION

Nutrients are necessary for seaweed growth. Recently, the widespread implementation of water purification has progressed and the concentration of nutrients in coastal areas has tended to decrease. For seaweed growth, it is difficult to increase the nutrient concentration in coastal areas because a large amount of fertilization is required. Therefore, technologies were developed to provide the specific nutritional components that are essential for seaweed growth on the substratum. An alternative is a concrete substrate (Ion Culture plate) mixed with water-soluble glasses that gradually elutes divalent iron ions that promote the growth of seaweed (Hirose et al. 1999).

In addition, concrete that elutes amino acids has been developed. Arginine, one of the essential amino acids, has the characteristic of slowly leaching into the water from the concrete surface of the substratum for a long time without damaging the strength of the concrete. A considerable amount of microalgae tends to settle on the concrete substratum containing arginine. According to experiments (Nishimura et al. 2014), microalgae was five to ten times more abundant on arginine-leaching concrete compared to concrete lacking in arginine (Figure 9.7). Since small seaweeds serve as prey for herbivorous animals, it is expected that there will be an increase in the probability of survival of large seaweeds as herbivores prefer to feed on microalgae. An actual *in-situ* marine experiment was conducted using arginine-leaching concrete. The purpose was to add an environmental function to the wave-dissipating blocks deployed as a breakwater in the Port of Wajima, facing the open sea in the Sea of Japan. In this experiment, the desired results of using the arginine-containing concrete to promote seaweed growth, and subsequently create a habitat for various organisms, were confirmed. The results of the Port of Wajima Project (Yoshizuka et al. 2018) were evaluated and received the "Certificate of Recognition" by PIANC (World Association for Waterborne Transport Infrastructure) Working with Nature in 2018.

DEPLOYING SUBSTRATA

Seaweeds (such as kelp) tend to settle on convex surfaces of the substratum. Since the spores of these seaweeds are as small as several μm, it would be relatively easy for these spores to survive if there were projections where the substratum was exposed to waves or currents so that the spores will not become buried in the surrounding mud. Also, it is more likely that a dispersed spore will adhere to a substrate if the surface is irregular (i.e., the edge effect). In recognition of this attribute of the spores, devices have been designed to provide convex surfaces on the blocks. Floating mud often accumulates on artificial substrates. In these areas inverted trapezoidal protrusions (i.e., the Kelp Knob, Figure 9.8) can be installed so that floating mud does not accumulate. As a result, the settlement of kelp species is enhanced on the knob (Terawaki 1988, Shibata et al. 2015)

FIGURE 9.7 Environmentally active concrete (i.e., concrete mixed with amino acids) has a greater concentration of microalgae on its surface.

FIGURE 9.8 *Ecklonia cava* growing on Kelp Knob where it is hard for mud to accumulate.

Typical examples of deployed substrates to improve seaweed survival are grooved blocks and blocks made of porous concrete. The edges of both sides of the grooves provide a substrate where seaweed is likely to settle.

Porous concrete has coarse surfaces, as it is made without sand. It is also called No-Fines Concrete (NFC) and has continuous voids. Porous plates made of ceramics and shells have a greater roughness and a larger surface area than ordinary concrete structures, and therefore have the characteristics on which the spores of seaweeds are likely to settle.

Porous concrete plates made with ceramics can provide a complex surface structure to facilitate the settlement of kelp. Also, this surface makes it more difficult to remove established seaweed from the substratum (Figure 9.9). This porous concrete plate is made of recycled roof tiles, natural fibers, iron powder, and cement. The porosity is about 25% and the shape of the plate can be easily molded. The porous plate has a 20 mm diameter hole in the center, and can be attached to any block using chemical anchors or bolts.

Plates composed of seashells are porous and effective in facilitating the settlement of seaweed, because they inhibit the deposition of mud. Adhesion of seaweed on these plates has been observed for more than 10 years after deployment. The plate material apparently does not adversely affect the environment (Figure 9.10).

A hemispherical, porous substratum with a diameter of 50 cm has also been developed (Figure 9.11). Porous concrete has a complex surface and creates a large surface area compared to a flat surface. Porous concrete is permeable to water which reduces mud accumulation and pore

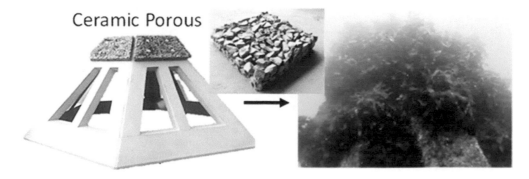

FIGURE 9.9 Artificial reef with ceramic porous plates.

FIGURE 9.10   Fish reef with a porous plate using shells.

FIGURE 9.11   Artificial reef with hemispherical, porous substrata.

clogging. It is not high in structural strength so it is placed on a stronger (less porous) concrete block to enhance its durability as a seaweed substrate.

Since it is difficult to form complex protrusions on concrete blocks, cast-iron substrata have also been developed (Figure 9.12). Surface texture with fine roughness is highly effective at facilitating seaweed spore settlement. The surface of this substratum can be renewed frequently, and it can be attached to concrete.

Since seaweeds change the composition of associated species (depending on the water depth), step-like blocks have been developed in anticipation of the formation of seaweed beds with various species assemblages (Figure 9.13). For example, a different assemblage of constituent species can be

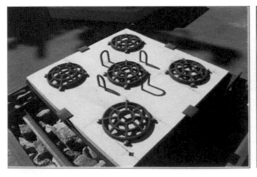

FIGURE 9.12   Cast iron seaweed substratum.

FIGURE 9.13 Fly ash concrete step-like blocks for seaweed adhesion.

expected at a seaweed bed composed of Sargassum. This block is called FA (fly ash) block because it is made of concrete that effectively uses fly ash, which is coal ash.

DISCUSSION

In conventional seaweed bed construction, the wave-resistant stability of the artificial substratum was regarded as important. Fishing ground structures that are stable against large waves create an environment in which not only seaweed, but also herbivorous animals can easily live. Immediately after deployment, there are few herbivorous animals associated with the substratum to serve as competitors, so seaweed grows on the new artificial substratum. However, after several years, there are many cases where the number of herbivorous animals (e.g., sea urchins) increase in the area and the standing crop of seaweed subsequently decreases. If the functional seaweed community is not maintained, the associated fishing grounds decline in productivity. Consequently, the constructed fishing grounds will be devastated and, if not repaired, will not be used by fishers.

From this perspective, the maintenance and management of the seaweed bed after its deployment is necessary to sustain the seaweed beds and the fisheries associated with them. Thus, it is important that the fishery facility is easy for fishermen to manage. The artificial substratum introduced here can limit the immigration of sea urchins and reduce invasion by herbivorous fish. Such substrata are expected to reduce the impact of management measures such as controlling sea urchin density by fishers. However, reducing the distance between artificial substrata results in an increase in habitat for herbivores. Since sea urchins cannot move on the sand, installing these facilities on the sand bottom can reduce the immigration of sea urchins to these structures. If sea urchins enter the seaweed areas, divers can readily see them and the sea urchins can be easily removed. On many coasts, natural reefs are found to depths of about 10m, while the sand bottom tends to occur deeper than 10 m. Therefore, installing artificial substrata on the sand bottom (Figure 9.14 below) facilitates fishery-based remedial action. Since the distance between artificial substrata is about 3 m, sea urchins cannot live between the substrata. By transplanting seaweed onto the substratum, the core of the seaweed bed is formed. Maintenance is enhanced because sea urchins do not easily cross the substratum between structures.

After the establishment of the structures, it is important to thoroughly remove sea urchins and herbivorous fish from the natural rock reef to restore the seaweed bed. Since the seeds are expected to be supplied from seaweeds on the artificial substratum, the distance between the natural reef and the artificial structures need not be increased. For future research efforts, it will be necessary to estimate the scattering distance of the seeds by the seaweed. Such a concept has been recognized by local governments, and designed artificial structures are currently being deployed and monitored in two locations in Nagasaki Prefecture.

As indicated above, cooperation between construction efforts and fisheries regulations are important. How to specifically plan and conduct these activities in Japan is currently at the trial-and-error

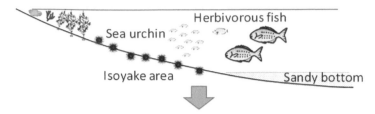

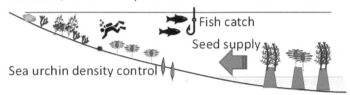

FIGURE 9.14 Conceptual diagram of recovering seaweed beds from sea urchin and fish damage (above, natural area with no additional artificial substrates; below, areas with artificial substrates added).

level. Under the initiative of the Japan Fisheries Agency, local governments nationwide are in the process of formulating a "Seaweed Bed Vision." A vision that considers excellent structure counter-measures and non-structure countermeasures that effectively use them is desirable.

This chapter introduced isoyake recovery technology using artificial substrata in Japan. However, this introduction was limited to the technology of companies belonging to the groups previously indicated. The Japan Fisheries Agency is planning a third revision of the *Isoyake Recovery Guidelines*. It is hoped that more advances in artificial structure design, construction, and deployment will be introduced, and specific interaction of these structures with fisheries regulations measures will be described.

REFERENCES

Hirose, N., A. Watanuki, S. Kawashima, M. Saiki, and S. Kitao. 1999. Seaweed bed formation an isoyake region in coastal areas of the japan sea, Hokkaido-Seaweed bed formation by removing sea urchin grazing pressure. *Proceedings of Civil Engineering in the Ocean* 15:165–170. (in Japanese)
Japan Fisheries Agency. 2007. *Isoyake Recovery Guidelines*. 213pp. (in Japanese)
Japan Fisheries Agency. 2015. *Isoyake Recovery Guidelines*. Revised edition. 199pp. (in Japanese)
Nishimura, H., R. Yamanaka, K. Sato, C. Tara, T. Nakanishi, and Y. Kozuki. 2014. Material performance of "Environmentally Active Concrete" containing amino acids in water. In *PIANC-World Congress*, San Francisco, CA, 1–17.
Shibata, S., A. Watanuki, and H. Kuwahara. 2015. Techniques and problems for construction of seaweed bed with the artificial substrate. *Fishery Engineering* 51(3):263–269. (in Japanese)
Terawaki, T. 1988. Preliminary study for creation of kelp forest. II. On relation between surface form of artificial algal substratum and colonization of young Ecklonia cava plant. CRIEPI Report, U88037, 26pp. (in Japanese)
Yoshizuka, N., H. Matsushita, T. Nakanishi, H. Nishimura and K. Oguma. 2018. Disaster prevention facilities and marine environment improvement effect. In *PIANC-World Congress*, Panama City, 1–13.

10 Why Do Japanese Fishermen Not Wear Life Jackets?
Answers Based on Interviews with Fishermen

Hideyuki Takahashi, Kenji Yasuda, and Kimiyasu Saeki

CONTENTS

ABSTRACT

Falling overboard is one of the major occupational hazards for fishermen in Japanese fisheries. Though wearing a life jacket is thought to be the most effective way to survive when a fisherman falls into the sea, many Japanese fishermen do not wear them. To understand why they do not wear life jackets, the authors observed three fishermen's work activities while wearing three types of life jackets designated by the authors (vest, collared, and waist-belt). The fishermen were then interviewed with regard to their perceptions of the jackets. Each life jacket was evaluated based on the fishermen's opinions on four aspects: ease in wearing or removal, mobility while wearing, risk of getting entangled in gear, and comfortability in hot weather. Two of the three fishermen preferred the waist-belt life jacket; however, the opinions were not the same among all fishermen. Because there are many different fisheries and aquaculture activities in Japan, the preference of fishermen for a particular type of life jacket may be different among them. These results indicate that we must continue a search to match life jackets with fishermen based on their actual work situations. Once concluded, the results of this effort must be disseminated to fishermen in all fisheries and regions in Japan.

INTRODUCTION

Fishing is thought to be one of the most dangerous occupations in Japan. Fishing has one of the three highest accident rates for all occupations, and it is only higher for occupations in forestry

and mining (Ministry of Health, Labour and Welfare of Japan 2018). Though the kinds of accidents are varied, falling overboard is one of the major occupational hazards fishermen face
(Ministry of Land, Infrastructure, Transport and Tourism of Japan 2017). It is paramount that
strict measures are required to decrease the incidence of falling-overboard accidents, because
they are directly connected to the danger of a fisherman's life. Once a fisherman falls overboard
into the sea, however, wearing a life jacket is thought to be the most effective way to increase the
chances of survival. The Japan Fisheries Agency reports that the mortality rate, due to falling
overboard, of fishermen who wore life jackets was 25%, nearly half of the rate of those who did
not wear them (52%), over the last 5 years (Japan Fisheries Agency 2018a). These data apparently show the effectiveness of wearing a life jacket when falling overboard. The percentage of
fishermen wearing life jackets during their work at sea, however, is only 35.4% (Japan Fisheries
Agency 2018b). Why do many Japanese fishermen not wear life jackets? One rational answer may
be that these jackets interfere their activities while at work at sea. The majority of fishermen's
tasks involve physical labor and life jackets may somewhat hinder their movements. Here, field
investigations were conducted by having several fishermen wear different life jackets. After their
experiences, these fishermen were interviewed about subsequent problems when conducting their
work. In this chapter, the results of this investigation are summarized and discussed with regard
to the possible ways to increase the number of fishermen wearing life jackets while working at
sea.

MATERIALS AND METHODS

INVESTIGATED SITES AND TYPES OF FISHERIES

Three fishermen, engaged in different fisheries or aquaculture, were interviewed between 2014 and
2018. Fishermen were named as Fisherman A, B, or C with each engaged in coastal otter trawling,
coastal gill netting, and pufferfish farming, respectively, as shown in Table 10.1. Each fisherman
worked on a boat that was less than 5 tons (Figure 10.1).

PROCEDURES

Before the interviews, each fisherman wore three different types of life jackets and worked at
least one cycle of their regular tasks separately on their boat (for instance, casting and hauling the
net and sorting the fish makes up one cycle for coastal otter trawling – a total of three tasks per
cycle). We accompanied the fishermen to allow us to observe and video-record their work. During
the interviews, they were asked about the comfort of the tested life jackets and the problems each
life jacket posed in their work. The answers from the fishermen were classified into four categories: (1) ease in wearing and removing, (2) mobility while wearing a life jacket, (3) risk of getting
entangled in the fish-net or equipment, and (4) comfortability in hot weather. Tested life jackets

TABLE 10.1

Profiles of Investigated Subjects

Fisherman	Site (city, prefecture)	Type of fishery	Main tasks investigated
A	Himeji, Hyogo	Coastal otter trawling	Net-casting and hauling. Sorting the fish
B	Echizen, Fukui	Coastal gill netting	Net-casting and hauling. Detaching fish from the net
C	Obama, Fukui	Pufferfish farming	Catching the fish. Tooth extraction. Releasing the fish.

FIGURE 10.1 Fishing boats of investigated fishermen: (A) coastal otter trawler, (B) coastal gill netter, and (C) working boat for pufferfish farming.

(the details are as follows) were evaluated as good, acceptable, or poor for each of the four evaluation categories.

LIFE JACKETS

The types of tested life jackets were: vest, collared, and waist-belt (Figure 10.2). The vest life jacket was composed of a firm, solid material to provide buoyancy. The other two life jackets were inflatable and provided buoyancy with the help of CO_2 filled in a chamber. All were certified models, meeting the safety standards prescribed by the Ministry of Land, Infrastructure, Transport and Tourism of Japan.

RESULTS

PROFILES OF THE FISHERMEN'S TASKS

Fisherman A (coastal otter trawling) (Figure 10.3, upper row): during the net-casting task, the cod end of the trawl net was manually thrown into the sea first, two otter boards were attached on both ends of the net, and then the warps were unrolled by operating the hydraulic net winch. In the net-hauling task, the warps were hauled until the otter boards appeared at the sea surface; the otter boards were then detached from the net, and the cod end was led to the port side and recovered

FIGURE 10.2   Tested life jackets.

on the bow deck. In the fish-sorting task, the fisherman kneeled on the deck and sorted the fish by species and size.

Fisherman B (coastal gill netting) (Figure 10.3, middle row): in the net-casting task, the fisherman manually cast the floating buoys, anchors, and ropes connected to the net at the beginning and the end of the set. The net itself was cast automatically into the sea by propelling the fishing boat. In the net-hauling task, the net was hauled by the electric net hauler, and the captured fish were detached from the net.

Fisherman C (pufferfish farming) (Figure 10.3, lower row): before the task, the fisherman positioned their boat alongside a fish cage. Pufferfish were captured by a scoop net and released into a tank on the boat. The teeth of the pufferfish were extracted with a pliers. This is done to ensure the fish in the fish cage do not attack each other. These pufferfish were then thrown into an adjacent fish cage.

FISHERMEN'S OPINIONS ON LIFE JACKETS

Overall opinions of the fishermen are summarized in Table 10.2. As a common point of agreement in all three cases, chiefly because it covered most of the upper body, the vest life jacket was evaluated as poor on comfortability in hot weather. Fishermen had various experiences with regard to comfort when using the other life jackets.

During coastal otter trawling (Fisherman A), the front side of the vest life jacket would get pushed upward by the fisherman's thigh while kneeling, interfering with his ability to sort the catch. Though the collared life jacket was considered acceptable, more time was required to adjust the fit and it was evaluated as less comfortable in hot weather than the waist-belt life jacket.

Net-casting task　　　　　　　Fish-sorting task

Fisherman A (Coastal otter trawling)

Net-casting task　　　　　　　Net-hauling task

Fisherman B (Coastal gill netting)

Catching the fish　　　　　　　Tooth extraction

Fisherman C (Pufferfish farming)

FIGURE 10.3　Each fisherman's working landscape.

During the coastal gill netting activity (Fisherman B), one of the belt-ends of the waist-belt life jacket got caught in the gill net during the net-hauling task. Also, the vest life jacket was easier to wear and remove than the other life jackets. On comfortability in hot weather, the evaluation for the jackets was the same as in Case A.

When pufferfish farming (Fisherman C), no remarkable difficulties were noted in using all three types of life jackets. However, comfortability in hot weather was evaluated as poor for all three life jackets. The collared life jacket caused discomfort around the neck, while the waist-belt life jacket caused discomfort in the lower part of the fisherman's body because it obstructed the upper open end of the salopettes (bibbed overalls), making the fisherman hot, humid, and uncomfortable.

TABLE 10.2
Summary of Investigation Results

Subject: Life-jacket type	Evaluation point	(1) Ease of wearing/ removal	(2) Mobility while wearing	(3) Risk of getting caught	(4) Comfortability in hot weather
Fisherman A:	Vest	Good	Poor	Good	Poor
Coastal otter	Collared	Acceptable	Good	Good	Acceptable
trawling	Waist	Good	Good	Good	Good
Fisherman B:	Vest	Good	Good	Acceptable	Poor
Coastal gill	Collared	Acceptable	Good	Acceptable	Good
netting	Waist	Acceptable	Good	Poor	Good
Fisherman C:	Vest	Good	Good	Good	Poor
Pufferfish	Collared	Good	Good	Good	Poor
farming	Waist	Good	Good	Good	Poor

DISCUSSIONS

The appropriate life jackets for fisheries or aquaculture were investigated through the experience of fishermen under the conditions of the three case studies, each of which allowed the fishermen to indicate preferences and opinions with regard to various activities performed. There are many types of fisheries and aquaculture in Japan, and fishermen should expect a different performance from their life jackets depending on their work activity and the type of life jacket being used. Detailed features of the tasks of each fishery or aquaculture must be considered to provide substantial information on the appropriate life jacket for fishermen.

Also, it is important to note that the appropriate life jacket for gill net fishermen should further be discussed because the gill net itself easily snags on various things. If the life jacket of a gill net fisherman can easily get snagged, it means the fisherman faces a greater risk of entanglement in the net and, subsequently, falling into the sea. Generally, a life jacket must be worn over all clothing to perform as expected. However, as an exception, gill net fishermen should probably wear non-inflatable (i.e., solid) life jackets inside their raincoats. The solid-type life jacket has intrinsic buoyancy, which will be effective even if it is worn inside garments. Furthermore, the shapes of life jackets need to be improved to make it difficult for fishermen to get entangled in their gill nets. Efforts to improve life jacket designs are anticipated from life jacket manufacturers.

Three typical types of life jackets were examined in this study; however, there are variations in each type of life jacket according to models and manufacturers. It is desirable to evaluate these variations and to select a variation that is suitable according to each fisherman's specific activities. Specially the waist-belt type has many variations in the shapes and functions of air chambers. In some part of waist-belt type models, the wearer needs to operate it to float in the correct way when falling overboard occurs. An air-contained buoyant type which can also be suitable for fishermen has a similar appearance to the solid type but uses soft and lighter air enclosures as its buoyancy body. MLIT requires the use of life jackets with the Sakura mark (Figure 10.4), which shows the life jacket is the certified model of safety standard prescribed by MLIT. It is necessary to choose a suitable life jacket from various types with Sakura mark.

In 2018, MLIT revised the applicable acts and ordinances which now legally oblige all passengers on small-sized vessels to wear life jackets (Ministry of Land, Infrastructure, Transport and Tourism of Japan 2017). Furthermore, violation points will be added for a captain who does not make the passengers wear life jackets, and his pilot's license will be suspended for up to 6 months due to the accumulation of the points from 2022. These initiatives are expected to drastically increase fishermen wearing their life jackets.

However, to further increase the number of fishermen wearing life jackets, efforts must be made to develop better life jackets that do not disturb fishermen's ability to function normally while conducting

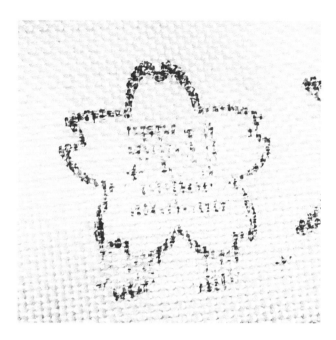

FIGURE 10.4 Example of Sakura mark stamped on a life jacket.

difficult and awkward tasks at sea. Moreover, it is important that information regarding the safety and comfort of each type of life jacket is disseminated to fishermen. In an effort to disseminate the results of our investigations (including the cases reported in this paper), they were compiled into a chart to help with choosing appropriate life jackets matching typical tasks in fisheries (Figure 10.5). The chart has been published on the Japan Fisheries Agency website (Japan Fisheries Agency 2018c).

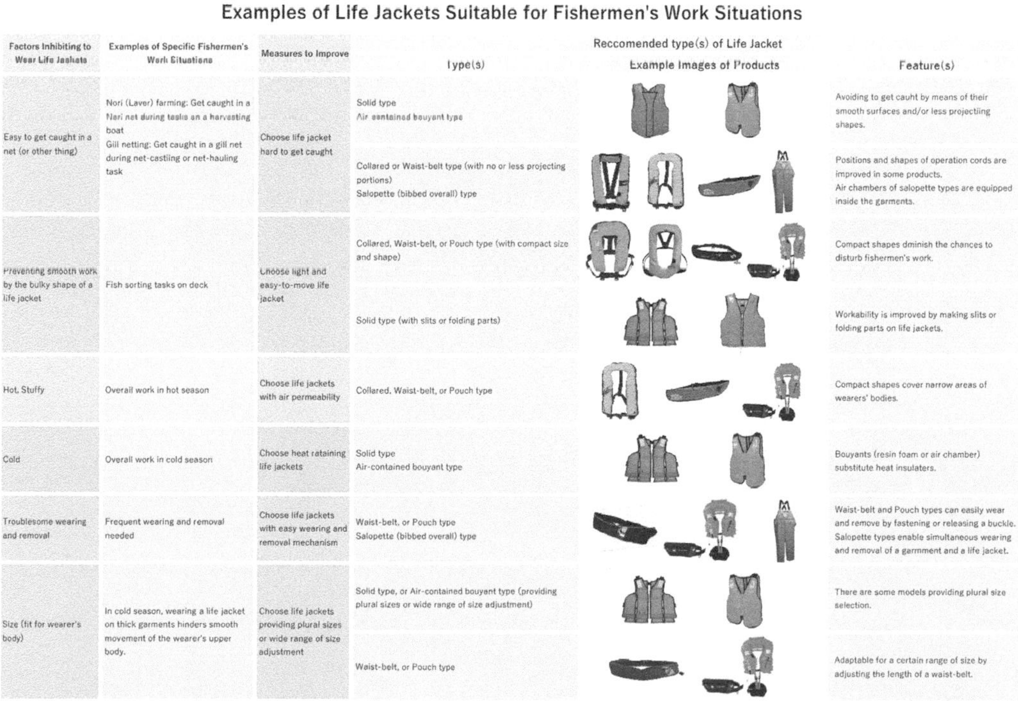

Examples of Life Jackets Suitable for Fishermen's Work Situations

Factors Inhibiting to Wear Life Jackets	Examples of Specific Fishermen's Work Situations	Measures to Improve	Recommended type(s) of Life Jacket		
			Type(s)	Example Images of Products	Feature(s)
Easy to get caught in a net (or other thing)	Nori (Laver) farming: Get caught in a Nori net during tasks on a harvesting boat. Gill netting: Get caught in a gill net during net-casting or net-hauling task	Choose life jacket hard to get caught	Solid type / Air-contained bouyant type		Avoiding to get cauht by means of their smooth surfaces and/or less projecting shapes.
			Collared or Waist-belt type (with no or less projecting portions) / Salopette (bibbed overall) type		Positions and shapes of operation cords are improved in some products. Air chambers of salopette types are equipped inside the garments.
Preventing smooth work by the bulky shape of a life jacket	Fish sorting tasks on deck	Choose light and easy-to-move life jacket	Collared, Waist-belt, or Pouch type (with compact size and shape)		Compact shapes diminsh the chances to disturb fishermen's work.
			Solid type (with slits or folding parts)		Workability is improved by making slits or folding parts on life jackets.
Hot, Stuffy	Overall work in hot season	Choose life jackets with air permeability	Collared, Waist-belt, or Pouch type		Compact shapes cover narrow areas of wearers' bodies.
Cold	Overall work in cold season	Choose heat retaining life jackets	Solid type / Air-contained bouyant type		Bouyants (resin foam or air chamber) substitute heat insulaters.
Troublesome wearing and removal	Frequent wearing and removal needed	Choose life jackets with easy wearing and removal mechanism	Waist-belt, or Pouch type / Salopette (bibbed overall) type		Waist-belt and Pouch types can easily wear and remove by fastening or releasing a buckle. Salopette types enable simultaneous wearing and removal of a garmment and a life jacket.
Size (fit for wearer's body)	In cold season, wearing a life jacket on thick garments hinders smooth movement of the wearer's upper body.	Choose life jackets providing plural sizes or wide range of size adjustment	Solid type, or Air-contained bouyant type (providing plural sizes or wide range of size adjustment)		There are some models providing plural sizes selection.
			Waist-belt, or Pouch type		Adaptable for a certain range of size by adjusting the length of a waist-belt.

FIGURE 10.5 Chart for choosing appropriate life jackets matching typical fishing tasks. (The original chart is written only in Japanese and the English translation was done by the authors.)

ACKNOWLEDGMENTS

This chapter summarizes part of the investigation implemented under the project for establishing a safe work environment for fishermen, funded by the Japan Fisheries Agency. We are grateful to the fishermen, and the fisheries staff of cooperative associations, local governments, and life jacket manufacturing companies, who supported our investigations. We are also thankful to Professor Shuji Hisamune (Kanagawa University) for his helpful advice.

REFERENCES

Japan Fisheries Agency. 2018a, Regarding the expansion of the duty of wearing life jackets, p. 2, (in Japanese) (https://www.jfa.maff.go.jp/j/kikaku/anzen.html#life, checked in February 13, 2020).

Japan Fisheries Agency. 2018b, Regarding the survey result of fishermen wearing like jackets, p. 1, (in Japanese) (https://www.jfa.maff.go.jp/j/kikaku/anzen.html#life, checked in February 13, 2020).

Japan Fisheries Agency. 2018c, Examples of life jackets matching work situations, p. 1, (in Japanese) (https://www.jfa.maff.go.jp/j/kikaku/anzen.html#life, checked in February 13, 2020).

Ministry of Health, Labour and Welfare of Japan. 2018, Trend of casualty rate per 1,000 persons per year by industry 2012–2018, (in Japanese) (https://anzeninfo.mhlw.go.jp/user/anzen/tok/anst00.htm, checked in February 13, 2020).

Ministry of Land, Infrastructure, Transport and Tourism of Japan. 2017, Implementation plan for prevention of seafarers' disasters, p. 5, 2017 (in Japanese) (https://www.mlit.go.jp/common/001175839.pdf, checked in June 14, 2019).

Ministry of Land, Infrastructure, Transport and Tourism of Japan. 2017, MLIT published the ministerial ordinance partially revising the Ordinance for Enforcement of the Law for Ships' Officers and Boats' Operators, (in Japanese) (https://www.mlit.go.jp/report/press/kaiji06_hh_000139.html, checked in October 11, 2019).

11 Habitat-Creation in the Sustainable Development of Marine Renewable Energy

Hideaki Nakata

CONTENTS

ABSTRACT

Marine renewable energy is widely recognized as an alternative energy source to mitigate the effects of climate change linked to CO_2 emissions, and there has been rapid expansion of offshore wind farms in Europe. The development of marine renewable energy is not without environmental concerns, including ecological costs and benefits of each project. Among the benefits are that it is highly likely that structures associated with marine renewable energy have the potential to act as artificial reefs, and possibly contribute to creation of new, hard-substrate habitats for a number of sessile and motile colonizing species. The habitat-creation potential, due to this artificial reef effect, could be a key issue of the sustainable development of marine renewable energy. This chapter presents a brief review of the existing knowledge on the reef effect, and indicates future directions for research. The findings from monitoring of the effects of "windmill artificial reefs" in the North Sea are summarized as a basis for considering research needs. Ecosystem-based approaches, reliable and quantitative assessment, and ecological consideration of the structure designs are recommended as future research tasks. Finally, the scientific basis of the ecological benefits from artificial reef effects is also discussed in relation to the possibility of co-locating marine renewable energy development with local fisheries.

INTRODUCTION

It is now widely recognized that there must be a paradigm shift in energy production from fossil fuels to alternative energy sources such as marine renewables if we are to mitigate the effects of climate change linked to CO_2 emissions. Primary among these renewable sources is the offshore wind energy sector, which has rapidly expanded in size and the number of installations in recent years. The potential to capture energy from waves or tidal currents has seen increasing interest, with pilot developments in a number of countries (Inger et al. 2009, Miller et al. 2013).

The technological development of marine renewable energy, however, is not without environmental concerns. The large scale of proposed marine renewable energy development will add to the existing human pressures on coastal ecosystems. Moreover, it is increasingly accepted that developments of any kind should only proceed if they are ecologically sustainable and will not reduce current or future delivery of ecosystem services (Gill 2005, Inger et al. 2009, Causon and Gill 2018). In this context, various frameworks to describe environmental effects and impacts of marine renewable energy development have been proposed (Boehlert and Gill 2010). Consequently, any ecological costs and benefits must be properly determined for each marine renewable energy project.

Among ecological benefits, it is highly likely that marine renewable energy installations have the potential to act as artificial reefs, and possibly contribute to the creation of new habitats for a number of sessile and motile colonizing marine species (Petersen and Malm 2006, Inger et al. 2009, Langhamer et al. 2009, Gill and Wilhelmsson 2019). Worldwide, artificial reefs are constructed and deployed in coastal waters to enhance fisheries and for habitat ecosystem restoration (Nakata 1995, Seaman 2007, Lima et al. 2019), and accumulated evidence suggests that artificial reefs generally hold greater fish densities and biomass, and possibly provide higher fish production than natural reefs (Power et al. 2003, Brickhill et al. 2005, Cresson et al. 2014). The novel, man-made structures constructed primarily for other purposes in the sea (e.g., wind turbines), also serve as habitat for fish and invertebrate assemblages. These have been defined as secondary artificial reefs because their function as artificial reefs was not the primary reason for their deployment (Pickering et al. 1998). Numerous wind turbine foundations (so-called windmill artificial reefs, Reubens et al. 2013a) have been deployed on the seabed of the North Sea and the Baltic Sea in Europe and, consequently, add a significant amount of artificial reef materials to the environment.

Given that marine renewable energy installations have the potential to create additional habitats by introducing artificial reefs and to attract marine organisms, including commercially valuable fish species, the overall effects of the marine renewable energy installations on the marine environment will serve as a positive attribute for the associated human communities. This is because these structures contribute to enhancing the cooperative relationship among stakeholders, such as fishermen, directly associated with a marine renewable energy project. The aims of this chapter are therefore to conduct a brief review of the research on the environmental effects of windmill artificial reefs in Europe, and to suggest future directions of the research related to habitat-creation potential as a key issue in the sustainable development of marine renewable energy.

TARGETED MONITORING OF ARTIFICIAL REEF EFFECTS

As the European Union's (EU) commitment to renewable energy is projected to grow to 20% of energy generation by 2020, the use of marine renewable energy from wind, wave, and tidal resources is increasing (Dannheim et al. 2019). Wind turbines, arranged in offshore wind farms, generate the bulk of the marine renewable energy in Europe. The most obvious habitat change coinciding with the construction of an offshore wind farm is undoubtedly the introduction of artificial hard substrata into the environment. The wind turbine foundations contribute to the creation of a new habitat for hard-substratum-associated epifauna in a predominantly soft-sediment environment, leading to an increase in biodiversity (De Mesel et al. 2015). The hard body-part matrices of some sessile biota also harbor macroinvertebrates that, in turn, contain potential food resources for associated fish (Reubens et al. 2011). The fish and sessile organisms associated with the turbine foundations may further contribute to increased benthic productivity (as measured by standing biomass), around windmill artificial reefs, possibly enhancing trophic cascades, which play a pivotal role in ecosystem function (Dannheim et al. 2019).

According to Rumes et al. (2013), benthic biomass in autumn at a Belgian offshore wind farm (i.e., Thorntonbank) showed a nearly 4,000-fold increase in biomass at the lower level of a turbine foundation between pre- and post-deployment. Considering the entire wind farm, there was a 14-fold increase in biomass observed at the turbine foundations. An increase of as much as 3% of the current estimated biomass in the Belgian part of the North Sea could be expected in the entire Belgian wind energy zone. Conversely, the species that settle on the artificial substrata will often comprise non-indigenous species. They could take advantage of the niche space offered by the introduction of new habitat as a "stepping-stone" to expand their population size (Adams et al. 2014, De Mesel et al. 2015). The windmill artificial reefs thus may also promote the establishment and spread of non-indigenous species, leading to unwanted negative effects in some instances. It should be further recognized that local biodiversity enhancement is countered by disturbance to the natural soft-sediment environment.

In Figure 11.1, a schematic overview of the most important reef effects influencing demersal fish production at windmill artificial reefs is given (Reubens et al. 2014). Several mechanisms influence fish production at windmill artificial reefs by: (1) providing additional food and increasing the feeding efficiency, as denoted with bold solid lines, (2) providing shelter from currents and predators/fisheries, as denoted with dotted lines, (3) providing a suitable habitat for settling and immigrating individuals, as denoted with thin solid lines, and (4) causing stress (e.g., noise emission by operational wind turbines, increased predation pressure), as denoted with dashed lines. All these mechanisms have an influence on the carrying capacity of the area immediately surrounding the reef. The construction of offshore wind farms therefore gives the potential for additional habitat creation, often regarded as compensation for habitats lost (Wilson and Elliott 2009). If appropriately managed, the emplacement of the reef (i.e., the turbine foundation) may benefit wider environment due to "spillover" effects (i.e., export of food energy beyond the immediate areas surrounding the reef).

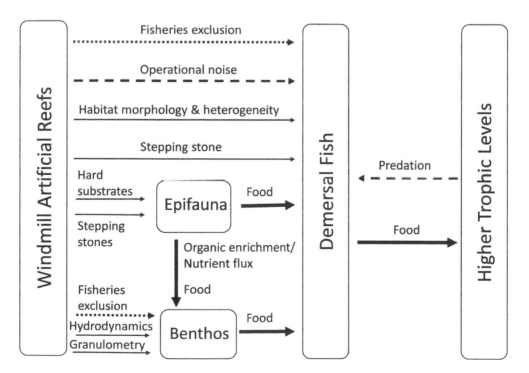

FIGURE 11.1 Schematic overview of the most important reef effects influencing demersal fish production at windmill artificial reefs. Each mechanism/process type is indicated by a line style (see the text for detail of the line style). (Modified from Figure 2 of Reubens et al. 2014).

As demonstrated in Figure 11.1, one of the marked changes in the local habitat is the attraction of fishes toward windmill artificial reefs. However, it is unclear whether or not these fishes are merely attracted to the structure, or are the result of increased production (i.e., growth or reproduction). With regard to this attraction/production issue, the behavioral ecology of benthopelagic fishes, mainly Atlantic cod and pouting, was investigated at the Thorntonbank Wind Farm in the Belgian part of the North Sea (Reubens et al. 2014). This investigation included data on the spatio-temporal variability of CPUE (catch per unit effort) for these two commercially important species at three different habitats (i.e., windmill artificial reefs, shipwrecks, and sandy bottoms). In addition, Reubens et al. (2014) collected biological data on: length, frequency distribution, diet, community structure, and movements to gain insights into the behavioral ecology associated with the fish assemblage and to determine if production occurred. It was revealed from these investigations that specific age groups of these benthopelagic fishes were seasonally attracted to the windmill artificial reefs (Reubens et al. 2013a, 2013b), and that these fish showed high site-fidelity and greater residence times around the wind turbines (Reubens et al. 2013c).

Figure 11.2 demonstrates the detection of Atlantic cod tagged by telemetry over time. In total, 22 individual cod were tagged with acoustic transmitters and released in the vicinity of two wind turbine foundations in the Thorntonbank offshore wind farm (for details of the methods, see Reubens et al. 2013c). As seen in the lower panel (Figure 11.2b), for detailed short-term detections, many fish were present near the wind turbine from May until October, with high individual detection rates. In addition, 97% of the detections were within a 50-m range of the wind turbine. These results indicate strong residency at the windmill artificial reef and distinct habitat selectivity by the fish. As determined from long-term detection data in the upper panel (Figure 11.2a), there were high numbers of fish present in summer–autumn, alternating with a period of very low numbers of fish present in winter. This indicated a clear seasonal association of fish presence near the windmill artificial reef. Analyses of stomach contents of pouting further revealed a high dietary preference for dominant epifaunal prey species such as *Jassa herdmani* (amphipoda) and *Pisidia longicornis* (decapoda) (Reubens et al. 2011, Reubens et al. 2013b). These species inhabited the hard substrate associated with the turbines, indicating that windmill artificial reefs could contribute to fish production on a local scale due to enhanced provision of resident food items on the turbines. Fish growth (i.e., size of fish) was actually observed throughout the period when the fishes were present. Moreover, the presence of juvenile Atlantic cod suggests their use of windmill artificial reefs as a nursery (Reubens et al. 2014).

In summary, while the fish community evidently responds to the presence of windmill artificial reefs, manifested as changing abundance, species assemblages, spatial and temporal distribution, and movement/migration, the overall question of whether windmill artificial reefs are ecologically beneficial for fish still remains open to debate. Increasing information about fish behavioral ecology, based on telemetry and bio-logging technologies, for example, evidently show that there could be different patterns in the use of windmill artificial reefs as a habitat, depending on fish species and their life cycle stages. Increased fish biomass could attract other organisms at higher trophic levels, such as marine mammals and seabirds, to the windmill artificial reefs (Scheidat et al. 2011, Vanermen et al. 2015). However, such a propagating reef effect at the ecosystem scale still remains unclear. Continuous monitoring and targeted research are further required to recognize more reliably positive effects of the windmill artificial reefs on the fish community and recruitment in the local fish population.

RECOMMENDATIONS FOR FUTURE RESEARCH ON THE HABITAT-CREATION POTENTIAL OF THE STRUCTURES ASSOCIATED WITH MARINE RENEWABLE ENERGY

ECOSYSTEM-BASED APPROACHES AND ECOLOGICAL MODELING

High potential for habitat-creation is being realized through well-designed monitoring at the windmill artificial reefs, as outlined above. However, the majority of existing data have been limited

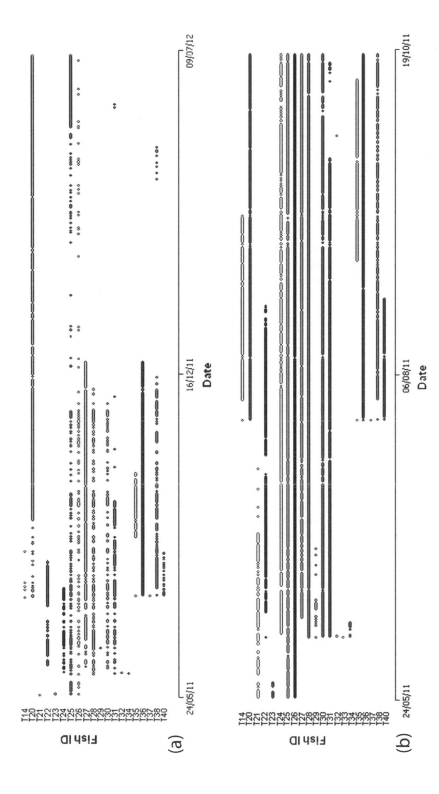

FIGURE 11.2 Overview of detections of all tagged Atlantic cod released from the vicinity of two wind turbine foundations in the offshore wind farm at Thorntonbank, Belgium. Each line represents the detections of one fish. Different shades of the lines (dark and light) distinguish the individual fish record. (a) Long-term detections from May 24, 2011 until July 9, 2012; (b) Detailed detections from May 24, 2011 until October 20, 2011. (Adapted from Figure 2 and Figure 4 of Reubens et al. 2013c.)

to single species at only a small number of installations. Critically, investigations will need to be conducted to determine if ecologically significant changes have occurred. Taking an integrated ecosystem view will provide a clear picture of knowledge gaps that relate to the cause–effect pathways of change and trophic interactions. These gaps can be filled by conducting targeted research (Lindeboom et al. 2015, Gill and Wilhelmsson 2019).

A recently published review article dealt with this issue (Dannheim et al. 2019). Focusing on benthic effects of offshore renewable energy development, this article proposed conceptual path diagrams to describe the cause–effect relationships of abiotic and biotic processes altered by the construction and operation of marine renewable energy facilities. Through such a systematic consideration of complicated ecological processes, it is possible to identify prominent knowledge gaps and future research needs. Additionally, this article emphasized the necessity of conducting hypothesis-driven research, in combination with integrative ecological modeling, on potential changes of ecological functioning through trophic cascades. Indeed, this approach is a promising way to apply ecosystem models to exploring artificial reef effects of marine renewable energy installations, and to improve the predictive capability of these deployments. This type of examination is still only in its infancy.

Recently, to simulate the reef effect inside an offshore wind farm with regard to trophic web functioning, an Ecopath ecosystem model composed of 37 compartments was developed to describe the situation before the offshore wind farm construction. Subsequently, an Ecosim projection over a 30-year span was performed after increasing the biomass of the targeted benthic and fish components to their accumulated maximum, presumed to be from the reef effects (Raoux et al. 2017). As a result, an overall increase in biomass at higher trophic levels and in detritus feeding groups was predicted by this model. Interestingly, however, any major trophic structural and functional shift due to the offshore wind farm construction was not detected. Conversely, a 3D ecosystem model describing carbon and nitrogen budgets, combined with a dynamic energy budget model for estimating blue mussel growth, was applied to quantify local effects of blue mussels colonizing turbine foundations at the Nysted offshore wind farm, Denmark (Maar et al. 2009). It was evident from the model output that ecosystem dynamics in the vicinity of the foundations were driven by the high biomass of blue mussels and their associated effects owing to their ingestion of micro-plankton, excretion of ammonium, and egestion of fecal pellets. Further, Adams et al. (2014) applied coupled biological and hydrodynamic models to investigate the stepping-stone effect of offshore renewable energy structures off the coast of southwest Scotland. Particle tracks during larval dispersion, simulated by a hydrodynamic model, clearly showed that the addition of marine renewable energy structures as novel offshore habitats enabled the particles to settle at new sites on the Scottish coast and allowed cross-channel invasion. Based on the probabilities suggested by the models, it would be reasonable to monitor for invasive species that have a high invasion potential at key sites.

A serious limitation of ecosystem modeling is the lack of sufficient data for model inputs and opportunities for model validation. Monitoring impacts and benefits predicted by numerical models will greatly contribute to model validation and improvement of these predictive models.

METHODOLOGIES FOR RELIABLE AND QUANTITATIVE ASSESSMENT

In the environmental impact assessment of marine renewable energy installations, it has been demonstrated that the evidence to assess the impacts, both positive and negative, remains poor. This is largely because of methodological weaknesses, including that much of the data were derived from short-term studies. Particularly worrying are the lack of both replication and appropriate baseline comparisons (Inger et al. 2009). With regard to this matter, it should be noticed that the highest estimate of secondary production per unit area of seabed was reported off the California coast (Claisse et al. 2014). This study was conducted over a long-term for 5–15 years (between 1995–2011) with concurrent surveys conducted at the oil and gas platforms and nearby natural reefs as a baseline for comparison. In these surveys, statistically significant high rates of total production were observed at the platforms, compared to the natural reefs. The principal reason for this difference was considered to be high levels of recruitment, possibly due to the large increase in the availability of complex and structural habitats distributed throughout the water column.

The availability of quantitative information on changes in fish abundance, population size, and productivity is still limited and mostly uncertain. Consequently, it is desirable to take a precautionary approach that includes mapping of impact sensitivity and cautious siting from an ecological viewpoint when planning marine renewable energy installations. Also, long-term monitoring of the artificial reef effect after the submersion of marine renewable energy structures will be indispensable in providing the empirical evidence needed for more reliable and quantitative assessments. In addition, revised and updated information, based on those monitoring data obtained from currently deployed marine renewable energy structures can provide feedback into design and deployment decisions regarding marine renewable energy deployments.

ECOLOGICAL CONSIDERATIONS IN THE DESIGN OF MARINE RENEWABLE ENERGY STRUCTURES

The extent to which marine renewable energy structures attract marine life may largely be affected by the design and structural complexity of the turbines and their foundations, which enhance their functions as shelter and feeding areas for fish. Although marine renewable energy structures are not designed with fish attraction or productivity in mind, appropriate design and siting should be required to enhance the habitat-creation potential of the marine renewable energy installations. Petersen and Malm (2006), however, indicated that there were still large knowledge gaps on how to design a windmill footing for specific biological purposes. They proposed that research was needed into an understanding of spatial scales including an examination of the interaction of the various types of windmill footings and the surrounding natural community. They discussed the reef design in more detail at three different scales: the microscale, involving material, texture, and heterogeneity of the foundation material; the mesoscale involving the revetments and various types of scour protection; and the macroscale involving the structure, sizes, and spacing of patches of habitats at the level of the entire offshore wind farm. In this regard, it may be important to note that offshore wind farms differ from other structures in that modification of the local environment spans multiple structures. In other words, an offshore wind farm represents a network of interconnected smaller artificial reefs, rather than a single large reef. Therefore, the reef effects need to be estimated cumulatively across the entire footprint of an offshore wind farm (Causon and Gill 2018).

In addition to the nature of the reef created, the reef effect and its productivity is heavily dependent on the characteristics of the location. Location features include the distance to other reef or hard substrate communities and the habitat requirements of the native key species surrounding the reef (Langhamer 2012, Petersen and Malm 2006). For maximum benefits of the reef effect, timing of the deployment of offshore foundations is also important. The foundations should be laid prior to the main larval settlement period of target species to allow an early start to colonization. The optimal orientation of the foundation and specific placement of bed material is another factor to be considered. The site-specific features will create a variety of hydrodynamic conditions, thus creating a range of micro-niches available for colonization (Wilson and Elliott 2009).

Lacroix and Pioch (2011) further proposed the eco-design of wind turbine foundations to create fish habitats. Their discussion considered the long-term resilience of coastal ecosystems with an increased bio-oriented complexity to engineered structures under a common eco-engineering vision. With careful and elaborate eco-design from the earliest planning stages, there is potential for actual habitat-creation to occur. This could have far-reaching benefits for both the local and regional environment, as well as potentially improving local fisheries (Wilson and Elliott 2009). Interdisciplinary research on the eco-design of marine renewable energy structures should, therefore, be promoted in collaboration with fisheries ecologists and engineers. This collaboration will greatly contribute to the establishment of advanced habitat technology for enhancing reef effects of marine renewable energy installations.

CONCLUDING REMARKS

Scientific Basis for the Potential Ecological Benefits from Artificial Reef Effects

Marine renewable energy installations provide marine organisms with new hard-substrate habitats for colonization, thus functioning as artificial reefs. The artificial reef effect is most important when considering the benefits of the marine renewable energy installations as described above; however, this feature of marine renewable energy has received very little attention in environmental impact assessments to date (Petersen and Malm 2006). To provide clear evidence of the potential ecological benefits, a scientific basis for the artificial reef effects should be demonstrated. Since artificial reefs have long been a major subject in Fisheries Engineering (Grove et al. 1991, Ito 2011), much could be learned from the existing knowledge of artificial reefs on how to strengthen the reef effects of marine renewable energy applications.

According to the recent overview by Lima et al. (2019), publications on artificial reefs have been increasing worldwide for several decades, and ecological studies continue as the main subject in these publications. Wind farms and other structures are actually considered as materials/structures used to create artificial reefs. The attraction/production debate, for example, in the artificial reef studies (Bohnsack 1989, Cresson et al. 2014) may be common to the question about the ecological function of marine renewable energy structures. Despite a wealth of literature regarding the ecological function of artificial reefs, no decisive evidence has been provided to solve the attraction/production debate (Brickhill et al. 2005). In recent years, however, studies of isotopic ratios have elucidated food-web functioning of artificial reefs (Cresson et al. 2014). These studies have revealed that attraction and production are two extremes of a range of contrasting patterns. Benthic sedentary species dominate at artificial reefs through massive fish production, while the initial predominance of pelagic, mobile species with low affinity for the reefs is explained by fish attraction (Cresson et al. 2019). This indicates that functional attributes of the associated assemblage, such as trophic traits, possibly modulate fish species responses to the deployed artificial structure. Additionally, the reef design could be critical to establishing the functional role of a reef. Interestingly, a detailed comparison between natural and artificial reefs off Australia suggested that reef design is an important determinant of species diversity and assemblage structure, while the location of the reef has a more significant influence on fish abundance (Komyakova et al. 2019). Updated information on the effect of reef location, design, size, and configuration on fish abundance and trophic functioning will provide valuable information to support the technological development and appropriate site selection of marine renewable energy installations.

Possibility of Co-Location of Marine Renewable Energy Development with Local Fisheries

The reef effects could also be beneficial to co-location of marine renewable energy development with the local fisheries. As an example, there is some evidence from ecological surveys in the UK that new hard substrate habitats created by offshore wind farm infrastructure support populations of the commercially important decapods, brown crab and the European lobster, providing the possibility of co-locating offshore wind farm with the local crab and lobster fisheries (Hooper and Austen 2014). Hence, in-depth interviews with the fishermen, conducted by Alexander et al. (2013), suggest that fishermen recognize the potential benefits of the reef effects. However, uncertainty remains, due to the limited scientific information currently available, which is also reflected in the uncertainty of the fishermen. The clear evidence that the reef effects of marine renewable energy installations could balance any loss of productive fishing grounds is apparently needed. In addition, the issue of scale is a possible limitation on the potential benefits to fishermen, because the footprint of the foundations, and hence the artificial reef effects, covered a very small fraction of that total site (Hooper et al. 2015). In the future, advanced technology may result in fewer, larger turbines, thus increasing the footprint of each foundation but reducing that of the wind farm as a whole (Wilson et al. 2010). Interdisciplinary and comprehensive studies on the reef effects, including the socio-economic aspects, are required for the realization of successful co-location between the renewable energy and fisheries sectors.

For avoiding unnecessary conflicts with the local fisheries sector as a key stakeholder, it is desirable to involve fishermen and other stakeholders in all stages of marine renewable energy development, including planning stages for structure design and siting, baseline surveys, impact assessment, and post-construction monitoring for adaptive environmental management (Inger et al. 2009). Further, enhanced fisheries management within the area of marine renewable energy installations, as a *de facto* marine protected area, would be critical to maximizing ecological benefits from the reef effect, as in the case of the artificial reef placements (Wilson et al. 2010, Reubens et al. 2014, Power et al. 2003).

Finally, offshore wind farms in Europe are largely deployed in areas with a subarctic climate with lower water temperature compared to the regions with temperate and subtropical climates. This suggests that aggregation of sessile biota on the offshore foundations could be more prominent in the temperate and subtropical regions. This includes regions of east and southeast Asian countries, as opposed to Europe, where research on the artificial reef effects has been concentrated. When marine renewable energy installations are planned in those regions with higher water temperatures, particular concern will be required over this issue. According to Nakata (1995), most of the coastal habitats are suffering from severe environmental degradation as a result of past human exploitation with less concern for the ecological cost. In these areas, there is a growing interest in enhancing and restoring the ecological quality of the coastal water. In this connection, marine renewable energy development could have the capacity to enhance biodiversity in degraded marine habitats and to restore reduced fisheries production through the artificial reef effects. An efficient habitat-creation technology should be established in harmony with the ecology, leading to a good practice as a basis for the longer-term future of marine renewable energy development.

REFERENCES

Adams, T.P., R.G. Miller, D. Aleynik, et al. 2014. Offshore marine renewable energy devices as stepping stones across biogeographical boundaries. *J. Appl. Ecol.* 51:330–338.

Alexander, K.A., T. Potts, and T.A. Wilding. 2013. Marine renewable energy and Scottish west coast fishers: exploring impacts, opportunities and potential mitigation. *Ocean Coast. Manag.* 75:1–10.

Boehlert, G. W., and A.B. Gill. 2010. Environmental and ecological effects of ocean renewable energy development: a current synthesis. *Oceanography* 23:68–81.

Bohnsack, J.A. 1989. Are high densities of fishes at artificial reefs the result of habitat limitation or behavioral preference? *Bull. Mar. Sci.* 44:631–645.

Brickhill, M.J., S.Y. Lee, and R.M. Connolly. 2005. Fishes associated with artificial reefs: attributing changes to attraction or production using novel approaches. *J. Fish. Biol.* 67:53–71.

Causon, P.D., and A.B. Gill. 2018. Linking ecosystem services with epibenthic biodiversity change following installation of offshore wind farms. *Environ. Sci. Policy* 89:340–347.

Claisse, J.T., D.J. Pondella, and M. Love, et al. 2014. Oil platforms off California are among the most productive marine fish habitats globally. *PNAS* 111:1542–15467.

Cresson, P., L.L. Direach, E. Rouanet, E. Goverville, P. Astruch, M. Ourgaud, and M. Harmelin-Vivien. 2019. Functional traits unravel temporal changes in fish biomass production on artificial reefs. *Mar. Environ. Res.* 145:137–146.

Cresson, P., S. Rutton, and M. Harmelin-Vivien. 2014. Artificial reefs do increase secondary biomass production: mechanisms evidenced by stable isotopes. *Mar. Ecol. Prog. Ser.* 509:15–26.

Dannheim, J., et al. 2019. Benthic effects of offshore renewables: identification of knowledge gaps and urgently needed research. *ICES J. Mar. Sci.* doi: 10.1093/icesjms/fsz018.

De Mesel, J., F. Kerckhof, and A. Norro, et al. 2015. Succession and seasonal dynamics of the epifauna community on offshore wind farm foundations and their role as stepping stones for non-indigenous species. *Hydrobiologia* 756:37–50.

Gill, A.B. 2005. Offshore renewable energy: ecological implications of generating electricity in the coastal zone. *J. Appl. Ecol.* 42:605–615.

Gill, A.B., and D. Wilhelmsson. 2019. Fish. In *Wildlife and Wind Farms, Conflicts and Solutions, Volume 3 Offshore: Potential Effects*, ed. M.R. Perrow, 86–111. Exeter: Pelagic Publishing.

Grove, R.S., C.J. Sonu, and M. Nakamura. 1991. Design and engineering of manufactured habitats for fisheries enhancement. In *Artificial Habitats for Marine and Freshwater Fisheries*, eds. W. Seaman, Jr. and L.M. Sprague, 109–152. San Diego: Academic Press, Inc.

Hooper, T., M. Ashley, and M. Austen. 2015. Perceptions of fishers and developers on the co-location of offshore wind farms and decapod fisheries in the UK. *Mar. Policy* 61:16–22.

Hooper, T., and M. Austen. 2014. The co-location of offshore windfarms and decapod fisheries in the UK: constraints and opportunities. *Mar. Policy* 43:295–300.

Inger, R., M.J. Attrill, S. Bearshop, et al. 2009. Marine renewable energy: potential benefits to biodiversity? An urgent call for research. *J. Appl. Ecol.* 46:1145–1153.

Ito, Y. 2011. Creation of fishing ground: artificial reef and present conditions. *Fish. Eng.* 48:157–160.

Komyakova, V., D. Chamberlain, G.P. Jones, et al. 2019. Assessing the performance of artificial reefs as substitute habitat for temperate reef fishes: implications for reef design and placement. *Sci. Total Environ.* 668:139–152.

Lacroix, D., and S. Pioch. 2011. The multi-use in wind farm projects: more conflicts or a win-win opportunity? *Aquat. Living Resour.* 24:129–135.

Langhamer, O. 2012. Artificial reef effect in relation to offshore renewable energy conversion: state of the art. *Sci. World J.* 2012:1–8.

Langhamer, O., D. Wilhelmsson, and J. Engstrom. 2009. Artificial reef effect and fouling impacts on offshore wave power foundations and buoys – a pilot study. *Estuar. Coast. Shelf Sci.* 82:426–432.

Lima, J.S., I.R. Zalmon, and M. Love. 2019. Overview and trends of ecological and socioeconomic research on artificial reefs. *Mar. Environ. Res.* 145:81–96.

Lindeboom, H., S. Degraer, J. Dannheim, et al. 2015. Offshore wind park monitoring programmes, lessons learned and recommendation for the future. *Hydrobiologia* 756:169–180.

Maar, M., K. Bolding, J.K. Petersen, et al. 2009. Local effects of blue mussels around turbine foundations in an ecosystem model of Nysted off-shore wind farm, Denmark. *J. Sea Res.* 62:159–174.

Miller, R.G., Z.L. Hutchinson, A.K. Macleod, et al. 2013. Marine renewable energy development: assessing the Benthic Footprint at multiple scales. *Front. Ecol. Environ.* 11:433–440.

Nakata, H. 1995. Production enhancement and its implications for the restoration of marine biodiversity in the coastal waters of Japan. *FAO Fisheries Circular* 889:1–32.

Petersen, J.K., and T. Malm. 2006. Offshore windmill farms: threats to or possibilities for the marine environment. *Ambio* 35:75–80.

Pickering, H., D. Whitmarsh, and A. Jensen. 1998. Artificial reefs as a tool to aid rehabilitation of coastal ecosystems: investigating the potential. *Mar. Pollut. Bull.* 37:505–514.

Power, S.P., J.H. Grabowski, C.H. Peterson, et al. 2003. Estimating enhancement of fish production by offshore artificial reefs: uncertainty exhibited by divergent scenarios. *Mar. Ecol. Prog. Ser.* 264:265–277.

Raoux, A., S. Tecchio, J.-P. Pezy, et al. 2017. Benthic and fish aggregation inside an offshore wind farm: which effects on the trophic web functioning? *Ecol. Indic.* 72:33–46.

Reubens, J.T., U. Braeckman, J. Vanaverbeke, et al. 2013a. Aggregation at windmill artificial reefs: CPUE of Atlantic cod (*Gadus morhua*) and pouting (*Trisopterus luscus*) at different habitats in the Belgian part of the North Sea. *Fish. Res.* 139:28–34.

Reubens, J.T., S. Vandendriessche, A.N. Zenner, et al. 2013b. Offshore wind farms as productive sites or ecological traps for gadoid fishes? – Impact on growth, condition index and diet composition. *Mar. Environ. Res.* 90:66–74.

Reubens, J.T., F. Pasotti, S. Degraer, et al. 2013c. Residency, site fidelity and habitat use of Atlantic cod (*Gadus morhua*) at an offshore wind farm using acoustic telemetry. *Mar. Environ. Res.* 90:128–135.

Reubens, J.T., S. Degraer, and M. Vincx. 2011. Aggregation and feeding behaviour of pouting (*Trisopterus luscus*) at wind turbines in the Belgian part of the North Sea. *Fish. Res.* 108:223–227.

Reubens, J.T., S. Degraer, and M. Vincx. 2014. The ecology of benthopelagic fishes at offshore wind farms: a synthesis of 4 years of research. *Hydrobiologia* 727:121–136.

Rumes, B., D. Coates, I. De Mesel, et al. 2013. Does it really matter? Changes in species richness and biomass at different spatial scales. In *Environmental Impacts of Offshore Windfarms in the Belgian Part of the North Sea: Learning from the Past to Optimize Future Monitoring Programmes*, eds. S. Degraer, R. Brabant, and B. Rumes, 183–192. Brussels: Royal Belgian Institute of Natural Sciences.

Scheidat, M., J. Tougaard, S. Brasseur, et al. 2011. Harbour porpoises (*Phocoena phocoena*) and wind farms: a case study in the Dutch North Sea. *Environ. Res. Lett.* 6:025102.

Seaman, W. 2007. Artificial habitats and the restoration of degraded marine ecosystems and fisheries. *Hydrobiologia* 580:143–155.

Vanermen, N., T. Onkelinx, W. Courtens, et al. 2015. Seabird avoidance and attraction at an offshore wind farm in the Belgian part of the North Sea. *Hydrobiologia* 756:51–61.

Wilson, J.C., and M. Elliott. 2009. The habitat-creation potential of offshore wind farms. *Wind Energ.* 12:203–212.

Wilson, J.C., M. Elliott, N.D. Cutts, et al. 2010. Coastal and offshore wind energy generation: is it environmentally benign? *Energies* 3:1383–1422.

12 Offshore Wind Energy and the Fishing Industry in the Northeastern USA

Michael V. Pol and Kathryn H. Ford

CONTENTS

ABSTRACT

The offshore wind industry in the northeastern United States is on the verge of developing more than a 21 GW capacity in the next few years. The first offshore wind farm proposed in the region was never built, in part due to stakeholder lawsuits. Therefore, outreach to stakeholders has been emphasized as a necessity for newer projects. Fishing stakeholders represent economically and culturally important industries and are potentially heavily impacted by the development of offshore wind farms, so robust efforts have been made to engage fishing stakeholders through many forms of outreach, both formal and informal. Working groups organized by individual states have successfully identified several key issues from fishing stakeholders' perspectives, including access and safety, transit corridors, and impacts on marine resources. The fishing stakeholders have also influenced offshore wind development designs, research priorities, construction timing, and financial compensation arrangements. However, several conflicts remain unresolved and recent press reporting indicates that fishing stakeholders feel voiceless and powerless in the process of offshore wind energy development, potentially threatening development. Further outreach efforts should: establish a process to engage stakeholders at crucial offshore wind farm development steps such as siting and turbine placement, not just during permit review; establish the means and expectations for conflict resolution; build trust between developers and fishing stakeholders; and should seek help and advice from social sciences.

INTRODUCTION

Electrical generating capacity in the United States of America (USA) has declined due to the decreasing economic viability of coal-burning plants and the senescence of nuclear plants (ISO New England 2019). A possible renewable, low-carbon energy replacement is wind energy, as abundant wind resources are available in the middle of the country and along its coastlines (AWS Truepower 2012). Currently, infrastructure is lacking to distribute electricity generated by wind farms from the central USA to the demand centers along the coasts. The coastal waters of the shallow continental shelf off the Atlantic coast of the USA are close to the demand centers, and are targeted for renewable energy infrastructure development, primarily through the development of offshore wind farms.

Projects in shelf waters under federal jurisdiction in the USA require offshore wind developers to lease their proposed site from the Bureau of Ocean Energy Management (BOEM), the lead federal agency for offshore wind development. There is a competitive lease issuance process, after which developers begin to study their lease area, design a turbine array, and apply for permits to build their offshore wind farm. As of May 2019, 15 leases with a total area of 7,000 km^2 have been issued for offshore wind farm development on the Atlantic coast off the states of Massachusetts, Rhode Island, New York, New Jersey, Delaware, Maryland, Virginia, and North Carolina (Figure 12.1). As many as 3,000 individual turbines of 7 MW capacity or higher are expected to be built during the next decade, creating a total generating capacity of 21 GW (Bureau of Ocean Energy Management 2019).

The first project in the USA to lease an area to develop an offshore wind farm, Cape Wind, was located off the coast of Massachusetts. The permitting process for the 130-turbine, 450 MW project was completed, but construction faced lengthy delays due to lawsuits by stakeholders including non-governmental groups and fishing stakeholders. Ultimately, the project was not built as a result of these delays (Seelye 2017). In an effort to minimize delays in future projects, the federal government created Wind Energy Areas. Wind Energy Areas are designated through a public process that assesses ecological value and stakeholder conflict and attempts to identify regions where leasing for offshore wind development will have fewer adverse impacts. Developers are not required to lease

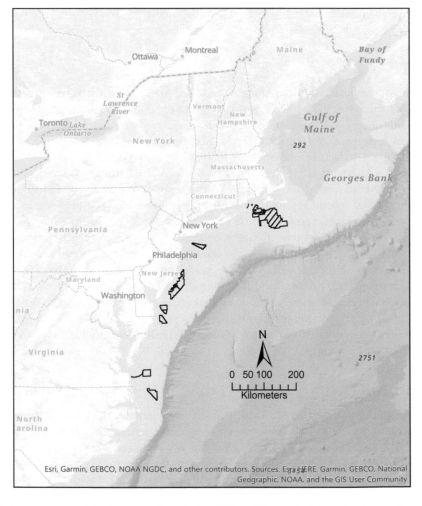

FIGURE 12.1   Areas leased (polygons outlined in black) for offshore wind farm development on the Atlantic coast of the USA in May 2019.

within a Wind Energy Area, but are encouraged to do so. To date, all leased areas are within a Wind Energy Area and all Wind Energy Areas are fully leased. Planning is underway for additional Wind Energy Areas off the coasts of New York, New Jersey, and further south. Planning is also ongoing for Wind Energy Areas northward in the Gulf of Maine and along the USA's west coast (Bureau of Ocean Energy Management 2019).

Continental shelf waters are used by a large variety of stakeholders including the US military, mineral interests, shippers, recreational boaters, and recreational and commercial fishermen (Nicholson et al. 2016). Therefore, despite efforts to minimize conflicts, Wind Energy Areas still overlap with other users. All of the current leased areas are partially or entirely open to recreational and commercial fishing and are anticipated to remain accessible to fishing after construction of offshore wind farms. Fishing has a high economic and cultural value in the USA and fishing stakeholders are a large and diverse group of both recreational and commercial fishing interests. Commercial fishing stakeholders include industries using a variety of commercial gears (otter trawls, scallop dredges, hydraulic clam dredges, bottom gill nets, and pots) to target midwater fish, groundfish, and shellfish (Pol and Carr 2000). Some of these commercial entities fish across the areas planned for wind development and land their catches in multiple states along the Atlantic coast. The recreational fishing stakeholders are primarily individual anglers using hook-and-line from a variety of platforms: private/rental boats, party/charter boats, and the shoreline. The majority of recreational fishing in the USA takes place on the Atlantic coast and is an important economic sector in all states on the Atlantic coast (National Marine Fisheries Service 2018). There is no known subsistence fishing in Wind Energy Areas or areas leased for wind energy development.

Federal law, through the US National Environmental Policy Act, requires the evaluation of potential socio-economic impacts of any development, including offshore wind farms. Due to the economic and cultural value of fishing in the USA, offshore wind farm development must by law account for adverse socio-economic impacts on fishing stakeholders and their home ports. Since offshore wind farms connect to shore through power cables that cross state, regional/county, and local boundaries, developers are also subject to permitting processes across multiple government jurisdictions and agencies. In general, outreach to stakeholders is required at each stage of the permitting process and by each governmental entity. Outreach methods include the solicitation of public comments on permitting documents, holding public meetings to discuss projects and project changes, and, in the case of offshore wind, requiring targeted communication with fishing stakeholders.

The purpose of outreach to stakeholders is to modify proposed offshore wind farms to minimize adverse effects on the environment and the stakeholders using the project site. For the purposes of this chapter, successful outreach is defined as creating engagement. Engagement connotes a level of stakeholder participation that results in modifications of offshore wind farms. Engagement is viewed by developers and regulators alike as improving the likelihood of successful permit application by decreasing opposition (Bureau of Ocean Energy Management 2015).

This chapter describes the types and methods of outreach used to communicate with fishing stakeholders and the resulting level of engagement of the fishing stakeholders in Massachusetts. We also assess how successful fishing stakeholder engagement has been in Massachusetts as of September 2019. Outreach is still ongoing and, therefore, different approaches are described and evaluated based on their immediate effectiveness.

MATERIALS AND METHODS

At the federal level, the permitting process for development mandates stakeholder outreach via public comment and also during subsequent appeals processes. In Massachusetts, reviews at the state, regional/county, and local stages of the permitting processes also provide opportunities for public comment. The typical process includes public notice of the action and solicitation of written comments at each stage. Public hearings and informational sessions may also be held to solicit comments (Bureau of Ocean Energy Management 2016) and are required for larger developments.

Other activities related to offshore wind farm development may be initiated that also require outreach via public notices, hearings, and comment periods. For example, in addition to permitting and public hearings for individual offshore wind farms, other public hearings were held by the US Coast Guard (USCG) to prepare for the possibility of altering vessel traffic in response to offshore wind farm construction and maintenance activities (United States Coast Guard 2015). At the federal level, permitting documents and the public comment period are organized by the lead federal agency of that action (e.g., BOEM or the USCG), whereas at the state, regional/county, and local levels the developer organizes the permitting documents and responses to comments, while the regulatory agency notifies the public and receives public comment.

Outreach requires identifying and communicating with affected stakeholders, which can be difficult for project developers who may be unfamiliar with the region. To address this challenge, BOEM developed guidelines for outreach to fishing industries affected by wind energy development. These guidelines are based on an outreach model from the United Kingdom (Fishing Liaison with Offshore Wind and Wet Renewables Group 2014), as modified by feedback gathered at public workshops with US fishing stakeholders (Bureau of Ocean Energy Management 2014). These guidelines recommend the development of communication plans that use Fisheries Liaisons and Fisheries Representatives. Liaisons are individuals hired by an offshore wind developer to be the developer's primary point of contact with fishing stakeholders. Fisheries Representatives are the fishing stakeholders' primary point of contact through which they can communicate concerns to the developer (Bureau of Ocean Energy Management 2015).

The Fisheries Liaisons employed by each offshore wind farm developer typically have prior experience, name recognition, and credibility in the region, and may have fished commercially in the past. Fisheries Representatives are either actively fishing or representatives of organizations which are often organized according to fishing gear types or species-targeted, such as groundfishing and scalloping. Liaisons and Representatives attend all outreach events and communicate with fishermen directly through conversations with individuals or small groups as a means to engage fishing stakeholders. Either through Liaisons or directly, developers use electronic communication media including email lists, social media, and text messages to alert fishermen of operations and changes in operations in lease areas, including geotechnical, fisheries, and other surveys.

Additional communication methods have evolved in each state to accommodate the needs and requests of developers, fishing stakeholders, or by government agencies. These methods vary in formality, structure, and function. The most formal process was developed in the state of Rhode Island, which legislated a process through its coastal zone management program that empowered an appointed board of commercial and recreational fishermen to speak for, and to negotiate with, developers on behalf of the fishing industry (McCann 2010). The Rhode Island Fisheries Advisory Board is formally organized, with defined membership that represents specific commercial gear types and recreational fishing sectors, as well as specific geographical regions. The Fisheries Advisory Board meets regularly, has a defined remit, posts minutes on a website, and some meetings are recorded.

In Massachusetts, the coastal zone management office and energy development agency created the Massachusetts Fisheries Working Group. This working group is informal, with open membership, no fixed terms of reference, and meetings on an as-needed basis. These meetings augment the BOEM planning process and provide a forum for fishing stakeholders to communicate directly with BOEM and offshore wind farm developers. Organizers disseminate information about upcoming meetings and public comment periods related to wind energy development to the Fisheries Working Group through an email list organized by the Massachusetts state agencies for this purpose. The Fisheries Working Group has met since 2009 primarily in in-person meetings. Since 2014, agendas have been available online (Massachusetts 2019).

The state of New York hired a mediator to serve as a point of contact for fishing stakeholders during the initial planning phases for locating offshore wind farms. In later phases of planning, the energy authority, marine fisheries agency, and the department of state invited specific fishing industry members to form the Fisheries Technical Working Group to provide guidance and advice on how to implement offshore wind farms. The group developed a framework document to define mission, objectives, and membership criteria (New York 2018). This group uses facilitators and remote attendance technologies, records its meetings, and produces minutes and work items.

For each of these working groups, government staff scientists and consultants generate information to be provided to the group to stimulate and to advance discussions on relevant topics. Along with Fisheries Liaisons and Fisheries Representatives, representatives from offshore wind farm developers and fishing stakeholders attend the meetings. These state working groups seek fishing stakeholder engagement on major topics, such as economic analyses and necessary research studies.

Regional fishery management organizations do not have a statutory role in the permitting process, but have also communicated information to fishing stakeholders via websites, email distributions, and by holding informational sessions. The Mid-Atlantic and New England Fishery Management Councils developed policies with respect to best management practices for offshore wind farm development and engagement of fishing industries (http://www.mafmc.org/northeast-off shore%20wind).

Private individuals have also communicated with fishing stakeholders. In at least one case, an interested party traveled on his own to Europe to speak to fishermen and to report back to US-based fishing stakeholders.

Through involvement in the Massachusetts and New York fisheries working groups, and participation in multiple task forces, research discussions, and individual conversations with fishermen, personal reflections are provided here on the success of stakeholder outreach efforts at generating engagement with fishing stakeholders to date. This discussion is focused on experience in Massachusetts related to the Massachusetts Wind Energy Area and the Fisheries Working Group.

RESULTS AND DISCUSSION

The minimum expectations for stakeholder outreach associated with the permitting of construction projects in the USA and suggested by BOEM appear to have been exceeded for offshore wind farm development. Government agencies, non-governmental organizations, and offshore wind farm developers have provided multiple communication pathways and outreach opportunities to engage fishing stakeholders both within and outside the formal permitting process.

Attendance and participation at meetings of the state-based working groups was, and remains, high, which can be interpreted as a validation of the value of the meetings and as a sign of engagement. Fishing stakeholders commonly respond to public comment periods associated with permitting. Written and verbal concerns and comments by these stakeholders have identified many topics at broad and specific levels. Concerns expressed by fishing stakeholders related to the design, construction, and operation of offshore wind farms focus on five main subjects: the specific location of Wind Energy Areas, leases, and offshore wind farms; obstruction of traditional transit routes; loss of fishing grounds due to inability or unwillingness to fish among the turbines; impacts on fishery resources; impacts on safety due to radar interference by turbines; and access for search-and-rescue personnel. The identification of these issues is helpful to government agencies and developers and indicates engagement by fishing stakeholders.

Further indication of engagement is that fishermen have identified many issues related to risk of impact from wind farm development on fishery resources, particularly the artificial reef effect, disruption of larval settlement, and the potential for interaction with fish sensitive to electromagnetic fields. These issues also concern scientists and managers (Massachusetts Division of Marine Fisheries 2018). Fishing stakeholders have been vocal advocates of the need to move beyond monitoring impacts within a single leased area and to meaningfully address cumulative impacts resulting from multiple offshore wind farms through the development of research plans to measure or mitigate cumulative impacts. Fishing stakeholder engagement is evident in their feedback on how to fund and manage programs. Fishing stakeholder concerns have led to additional funding for fisheries-related research for projects undertaken by BOEM, the USCG, offshore wind developers, and through state-supported research. Notably, a regional science organization, the Responsible Offshore Science Alliance (ROSA) was recently created. The Alliance is a collaboration among governments, developers, and the fishing industries intending to advance regional research and monitoring of fisheries and offshore wind interactions in federal waters.

Through engagement of fishing stakeholders, the need for appropriate spacing and corridors for fishing and transit to offshore fishing grounds was identified after leases had been issued to developers. The distance between turbine towers is relevant for fishing within offshore wind farms, since certain fisheries, particularly those using otter trawls, scallop dredges, or long-lines, have restricted mobility and need large areas for productive fishing. The first offshore wind farm developer held workshops to discuss fishing within offshore wind farms with fishing stakeholders in towns located closest to the offshore wind farm development. Fishing stakeholder engagement thus led to the first proposal for tower spacing and orientation. When the proposal was made public, fishing stakeholders that represented a larger geographic region were critical, and multiple conflicts resulted. Attempts were made to reach a consensus on the tower spacing and orientation. Information presented to fisheries working groups to facilitate discussion and resolution included scaled drawings that illustrated the relative scale of the gear to the towers (Figure 12.2), fishermen's chart plots, and federal vessel-tracking data. Regardless of the information available, fishing stakeholders counterproposed as great a distance as possible between towers. To accommodate this concern, developers altered the distance between towers to 0.7–1.0 nm (1.3–1.9 km).

The second offshore wind farm developer held workshops with fishing stakeholders that represented a broader, regional perspective. These workshops led to a second proposal using a minimum of 1 nm (1.9 km) spacing between towers and a standard north–south–east–west orientation. The developer that made the first proposal eventually agreed to use the tower spacing and orientation from the second proposal in later phases of wind farm development. This adjustment seemed to be a successful resolution of an issue identified through stakeholder engagement.

However, as discussions progressed, fishing stakeholders began to assert that fishing, especially with otter trawls and scallop dredges, could not occur within offshore wind farms regardless of the turbine spacing, due to radar interference and adverse weather conditions increasing the risk of allision. Additionally, the increased allision risk was linked to changes in vessel insurance policies that exclude coverage in offshore wind farms. The perceived level of risk of mobile gear interaction varied, but many stakeholders suggested that mobile gear fisheries simply would not be able to operate within offshore wind farms under any proposed configurations and distances.

A more nuanced conversation on optimal tower spacing was of interest to managers. If fishing is not possible in an offshore wind farm regardless of tower spacing, then the spacing could, in theory, be optimized to reduce the overall footprint of the development. Consequently, this arrangement could reduce habitat impacts and/or maximize power generation efficiency within the wind farm location. Maximization of efficiency could also result in less need for additional Wind Energy Areas and less area leased overall. However, fishing stakeholders also opposed closer spacing of the towers, presumably to protect potential fishing opportunities. A result was that several proposed wind developments have towers maximally spaced to be compatible with fishing. If fishing stakeholders are correct in asserting that they cannot fish within wind farms regardless of tower spacing, wider tower spacing may result in unnecessary costs to the wind farm developer and additional area occupied without achieving the goal of enabling fishing within. Nevertheless, accommodating fishing activities remains a priority, and the state of New York has provided funding to study mobile gear access in potential Wind Energy Areas. The conflict has also helped identify a need to better understand how alternative fishing gear could enable fishing in wind farms.

Fishing stakeholders also transit through the lease areas to access fishing grounds. The possible loss of this use of the areas was identified after leases were issued. To establish transit corridors, multiple meetings were held by the Massachusetts Fisheries Working Group to determine corridor widths and desired routes. These meetings led to the development of conflicting plans and

Turbine base

Typical otter trawl, Northeastern USA

Vessel length:	87 ft
Turbine distance:	0.78 nm
Turbine base:	70 ft x 70 ft
Depth:	30 fm
Wire ratio:	4.17:1
Wire out:	125 fm
Ground cables (with legs):	80 fm
Door spread:	47 fm
Net length:	240 ft
Total (bow to codend):	1540 ft

Note – gear dimensions vary with conditions, species, captain's preference and other factors

Drawn by Mike Pol – 13 August 2017

FIGURE 12.2 Scale drawing of a typical New England otter trawler within a proposed wind array of 0.7 nm (1.3 km) spacing.

recommendations and no consensus was achieved. The lack of resolution in part led to a formal analysis by the USCG, the federal agency responsible for navigation. The analysis, currently underway, provides additional stakeholder engagement opportunities and will determine if formal vessel routing measures are needed. Some meetings held for this study appeared sparsely attended by fishing stakeholders. Wind Energy Areas currently being planned in the New York Bight identified transit corridors by using vessel-tracking technology and through stakeholder engagement. These areas will not be leased for offshore wind farm development.

Fishing stakeholders were also influential in the development of compensatory mitigation plans that address the potential loss of revenue and fishing opportunity within offshore wind farms. Financial compensation arrangements were discussed at public and individual meetings with fishing stakeholders and a claims procedure for an existing 5-turbine offshore wind farm was developed with fishing stakeholders. The fishing stakeholders in both Rhode Island and Massachusetts were influential in assessing the quality and form of financial compensation agreements and their engagement modified the total compensation amount and compensation process.

Another example of engagement was the development of a specific sequence of construction activities for an offshore export cable that connects the offshore wind farm to the electrical grid on land. The cable was traversing squid grounds that have a short fishing season. In order to minimize the impact on the fishing activity, the cable-laying was timed to avoid the squid season.

As demonstrated above, fishing stakeholders have influenced the identification of research topics and the funding and initiation of research projects and financial compensation, the spacing and orientation of turbines within wind farms, the development of transit corridors, and the timing of construction. These examples suggest that outreach has been successful and has led to engagement, resulting in modifications to offshore wind farms to minimize adverse effects on fishing stakeholders. However, much conflict still exists. Both resolved and unresolved conflicts continue to be raised at meetings with fishing stakeholders after nearly ten years of discussion. Recently, fishing industry publications have described fishing stakeholders as being "drowned out," "shouted down," not being heard, and generally disrespected, thus concluding that outreach to fishing stakeholders has failed (Hathaway 2019, Parry 2019). This negative representation of interaction is inconsistent with the significant role outlined above.

The perception of lack of participation and engagement may delay development through resistance as previously occurred for Cape Wind. The first proposed offshore wind project, Vineyard Wind, south of Massachusetts, recently faced local and federal permit delays linked to disgruntled fishing stakeholders due to inadequate fisheries impact assessments.

Permitting delays due to unresolved issues and a perception of inadequate stakeholder engagement suggest that outreach and communication strategies are poorly managed, insufficient, or overconfident. More importantly, inadequate stakeholder engagement now and earlier may have resulted in flawed offshore wind farm siting and design decisions with far-reaching consequences.

The expectation by management agencies that a communication and outreach strategy will result in engagement by fishing stakeholders that will identify and resolve conflict and thus achieve co-existence of the two industries may be, in retrospect, naïve or an incorrect implicit assumption. Differing incentives of the different groups involved in outreach can lessen the degree of engagement. Managers and offshore wind developers want to develop renewable energy for policy and financial goals, but fishing stakeholders have less incentive to alter the status quo. Unspoken and unrecognized assumptions about the commonality of motivations and attitudes among fishing stakeholders, scientists, and managers and within these groups exist and act as obstacles to better understanding and conflict resolution. For example, co-existence of fisheries and wind energy development was declared at the Massachusetts Fisheries Working Group as a common goal, but not formally committed to or discussed by fishing stakeholders. Also, agencies and others commonly underestimate the complexity of attitudes among the fishing stakeholders with regard to change (Eayrs and Pol 2018). Engagement and conflict resolution might be more successful if the common benefits of renewable energy development were established and emphasized over individual or group impact. As required by statute and the usual practice, the framing of outreach efforts has concentrated on negative impacts to individual fishing operations. However, potential impacts to fishing operations by climate change may be far greater than offshore wind farm development. Replacing the recent loss of generating capacity by continuing previous energy policy via offshore oil and gas developments in this region is possible and would likely result in more severe impacts to fishing operations. These impacts are why, in part, there is a legislative mandate to replace energy supply to mitigate climate change brought on by greenhouse gas emissions. The potential impacts of maintaining the status quo and not building offshore wind farms have not been presented in the context of fishing stakeholder outreach for offshore wind farm development.

Other variables may have limited the effectiveness of outreach to fishing stakeholders. At the beginning of planning for offshore wind, the failure of the Cape Wind project introduced a natural skepticism that wind development would be successful, in turn limiting the participation and engagement of fishing stakeholders for subsequent offshore wind development activities. Despite multiple public meetings and formal comment periods over 5 years, after the first Wind Energy

Area was defined, many fishing stakeholders objected to the methods and outcomes. Fishing stakeholders indicated that fishing effort in the Wind Energy Area was underestimated due to known flaws of their self-reports and incomplete coverage by other monitoring systems. The insufficient stakeholder engagement during the siting of the Wind Energy Area consequently eroded trust during later actions. A marked increase in participation in outreach activities as specific potential threats to individual fishing stakeholders was observed once wind farm developers acquired leases and proposed turbine layouts. However, the distrust developed in the Wind Energy Area planning appeared to result in less cooperative conversations about how to address fishing stakeholder concerns through modifications of planned offshore wind farm developments.

The stakeholder outreach for the development of this new industry had other inherent flaws. Providing feedback on specific, consequential decisions in the development process has been difficult because of the timing of decisions. For example, for the Vineyard Wind project, the decision on spacing and orientation of the towers preceded any formal opportunities to comment or any requirement for stakeholder outreach. The Massachusetts Fisheries Working Group provided an opportunity to hear from developers and for the developers to interact with stakeholders. However, there was no process to discuss multiple layout options and the costs and benefits of those options, receive stakeholder feedback publicly, and come to a resolution.

A similar complicating factor in the permitting process also related to timing was the requirement that financial compensation for loss of fishing access be negotiated with fishing stakeholders prior to issuance of permits. Some permit issuance is on predetermined timelines related to financial incentives to the developers, with unclear capacity to modify the timelines. As a result, negotiations may have necessarily proceeded as rushed compromises rather than as thoughtful discussions. And, despite the level of involvement of the fishing stakeholders in the development of financial compensation agreements and a transparent procedure for establishing the level of compensation, the agreement approved for Vineyard Wind was perceived as unsatisfactory by some fishing stakeholders.

A clear shortcoming of the engagement process is that while conflicts are identified, a process for conflict resolution was never formalized or prioritized. For example, when the need for transit corridors was identified in the Massachusetts Wind Energy Area, subsequent meetings and discussions focused on producing a resolution were insufficiently planned and documented and lacked a systematic approach to solicit information from stakeholders and to resolve the issue. Failing to resolve conflicts such as this one has led to delays, further erosion of trust, and the need to revisit conflicts in subsequent meetings.

The reported perception and declaration of limited influence on decision-making by fishing stakeholders has itself increased distrust in the process, likely further reducing the effectiveness of future engagement efforts. Understanding and trust need to be re-established and maintained in outreach and engagement efforts by regulators and the wind energy developers with fishery stakeholders for successful development of wind energy. At the same time, the fishing stakeholders need to appreciate their influence and continue to engage constructively. The emergence of new organizations such as the Responsible Offshore Science Alliance is evidence of attempts at more constructive approaches to engagement.

CONCLUSIONS

Outreach and communication for the purposes of engagement could be judged to have been successful, but nevertheless, recent delays of the wind development process have resulted. Several explanations for perceived and real failures of the stakeholder engagement process for offshore wind are suggested. Among these are: unspoken assumptions about common goals and uncooperative approaches on the part of the stakeholders, poor timing of engagement on key decision points, and inadequate framing of discussions and conflict resolution by those responsible for outreach. Initially, limited stakeholder engagement was driven by the perception that offshore wind farm development was not immediate and it had a low probability of success. Offshore wind farm development is now viewed by many stakeholders as a present threat. To maximize the effectiveness of engagement with fishing stakeholders, outreach efforts should focus on 1) building or rebuilding trust through the creation of common goals by providing more information about the need for offshore wind and potential impacts of other energy developments; 2) building engagement around crucial development steps preceding permitting; and 3) identifying a clearer approach to conflict resolution. Conflict resolution may require specialized meeting structures and approaches. We recommend the use of social sciences in developing these strategies. To this end, a growing literature on the engagement of the public in wind energy development deserves fuller attention (e.g., Haggett 2011).

REFERENCES

AWS Truepower, LLC. 2012. Wind resource maps and data: methods and validation. https://aws-dewi.ul.com/assets/Wind-Resource-Maps-and-Data-Methods-and-Validation1.pdf

Bureau of Ocean Energy Management. 2014. Development of mitigation measures to address potential use conflicts between commercial wind energy lessees/grantees and commercial fishermen on the Atlantic outer continental shelf. BOEM OCS Study 2014-654. https://www.boem.gov/OCS-Study-BOEM-2014-654/

Bureau of Ocean Energy Management. 2015. Guidelines for providing information on fisheries social and economic conditions for renewable energy development on the Atlantic outer continental shelf pursuant to 30 CFR Part 585. https://www.boem.gov/Social-and-Economic-Conditions-Fishery-Communication-Guidelines/

Bureau of Ocean Energy Management. 2016. A citizen's guide to the Bureau of Ocean Energy Management's renewable energy authorization process. Washington, DC. 20 p.

Bureau of Ocean Energy Management. 2019. Lease and grant information. https://www.boem.gov/lease-and-grant-information/

Eayrs, S., and M. V. Pol. 2018. The myth of voluntary uptake of proven fishing gear: investigations into the challenges inspiring change in fisheries. *ICES Journal of Marine Science* 76(2):392–401.

Fishing Liaison with Offshore Wind and Wet Renewables Group. 2014. FLOWW Best Practice Guidance for Offshore Renewables Developments: Recommendations for Fisheries Liaison. https://www.thecrownestate.co.uk/media/1775/ei-km-in-pc-fishing-012014-floww-best-practice-guidance-for-offshore-renewables-developments-recommendations-for-fisheries-liaison.pdf

Haggett, C. 2011. Understanding public responses to offshore wind power. *Energy Policy* 39(2):503–510.

Hathaway, J. 2019. Whistling into the wind. *National Fisherman* 100(6):2.

ISO New England. 2019. *2019 Regional Electricity Outlook*. Holyoke, MA: ISO New England. 46 pp.

Massachusetts. 2019. Fisheries working group on offshore wind energy. https://www.mass.gov/service-details/fisheries-working-group-on-offshore-wind-energy

Massachusetts Division of Marine Fisheries. 2018. Management objectives and research priorities for fisheries in the Massachusetts and Rhode Island-Massachusetts offshore wind energy area. https://www.mass.gov/files/documents/2019/03/29/Management%20Objectives%20and%20Research%20Priorities%20for%20Offshore%20Wind%20and%20Fisheries%2011-5-18.pdf

McCann, J. 2010. Rhode Island ocean special management plan. Rhode Island Coastal Resources Management Council, Wakefield, Rhode Island. 3943 pp.

National Marine Fisheries Service. 2018. Fisheries of the United States, 2017. USA Department of Commerce, NOAA Current Fishery Statistics No. 2017. https://www.fisheries.noaa.gov/webdam/download/92995934

New York. 2018. Fisheries Technical Working Group Revised Framework. https://nyfisheriestwg.ene.com/Content/files/F-TWG%20Framework%20rev%20Nov2018.pdf

Nicholson, B., R. Getchell, and G. Fugate. 2016. Northeast ocean plan. Report to the National Ocean Council. 14 October 2016. 203 pp.

Parry, W. 2019. US fishermen demand to be heard on offshore wind energy projects. *Christian Science Monitor*, 17 September 2019. https://www.csmonitor.com/Business/2019/0917/US-fishermen-demand-to-be-heard-on-offshore-wind-energy-projects

Pol, M., and H. A. Carr. 2000. Overview of gear developments and trends in the New England commercial fishing industry. *Northeastern Naturalist* 7(4):329–337.

Seelye, K. Q. 2017. After 16 years, hopes for Cape Cod wind farm float away. *New York Times* 167(57,807):12.

United States Coast Guard. 2015. Atlantic coast port access route study. Final Report. Docket number USCG-2011-0351. 20 pp.

13 Hydrogen Fuel Cell and Battery Hybrid-Powered Fishing Vessels

Utilization of Marine Renewable Energy for Fisheries

Jun Miyoshi

CONTENTS

ABSTRACT

Here, a general arrangement and energy management strategy for a hydrogen fuel cell and battery hybrid-powered workboat for bluefin tuna farming in remote islands in Japan is presented. This boat will be supplied with hydrogen generated by electrolysis of water using offshore wind power in the future. First, an existing tuna farming workboat was examined for its actual general arrangement, hull form, and energy consumption in the field. Next, power unit components (e.g., fuel cell, battery, motor, and hydrogen tanks) were arranged based on Japanese regulations for a fuel cell boat, considering weight, trim, and the space in the ship hull. Third, several general arrangements were produced. Finally, energy supply and demand between wind power generators and hydrogen fuel cell/battery fishing vessels were estimated. A trial design of a hydrogen fuel cell/battery-powered fishing vessel using an existing vessel is introduced based on the 2018 hydrogen fuel cell ship safety guidelines in Japan, considering practical use in fisheries and usage of fuel cell stacks for automobiles.

INTRODUCTION

Renewable energy such as wind, wave, solar, and geothermal energy contribute to sustainable energy use and to the reduction of greenhouse gases (GHGs). As an alternative to fossil fuels, renewable energy sources can strengthen energy security. Especially for remote islands, renewable energy has possibilities not only to reduce GHGs, but also to supply essential energy for daily lives in cases when the power grid is shut down because of natural disasters. From the viewpoint of the electricity trade, renewable energy is a branding tool for environmental service organizations and companies that have a good awareness of environmental issues. Furthermore, new jobs and employment opportunities could be generated. A retail electric power market, which included renewable energy, by the Japan Electric Power Exchange (JEPX), was established in 2005 (Maekawa et al. 2018), and the electric market has been deregulated since April 2016 in Japan.

Concomitantly, the supply of natural energy such as solar and wind power at small offshore islands is unstable, depending on weather conditions. More importantly, natural energy generates less power compared to fossil fuels. As a method for storing and supplying electric energy, hydrogen and fuel cell stacks are effective. Storing hydrogen in tanks is desirable as a way to store energy and buffer the unpredictable delivery of electric energy, as well as to avoid overloading the capacities of power transmission lines (Obara 2019). Especially in island situations, continuous electric power generation from wind power is a viable prospect and storing this energy in hydrogen tanks is likely to be successful.

Fuel cell vehicles (FCV) using hydrogen have already been introduced to the market Barrett 2018), and hydrogen fuel cell/battery-powered ships have also been developed (e.g., the zero-emission ship (ZEM–SHIP) project and NEMO H2; van Biert et al. 2016). In Japan, Toda Corporation (2015) manufactured a trial hydrogen fuel cell-powered boat (an overhauled pleasure boat) in Goto City, Nagasaki Prefecture, in 2015 (Figure 13.1 and Figure 13.2). This boat was to be supplied with hydrogen generated by electrolysis of water using wind power installed on the offshore Fukue Island in Goto City Utsunomiya et al. 2015). Ode and his colleagues (Kifune et al. 2016, Ode and Shimizu 2019) developed a hydrogen fuel cell/battery-powered passenger boat which had indications of being highly practical in application. Hirata introduced a ship model in which a hydrogen fuel cell/battery system was constructed (Hirata and Miyazaki 2017). Additionally, Hirata and his colleagues (Hirata et al. 2019) investigated methods of preventing hydrogen leakage and demonstrated safe management of fuel cells while underway. The actual ship with a 60-kW fuel cell (30-kW fuel cell × 2) was launched and examined in the field. The results promoted the establishment of safety guidelines for fuel cell-powered boats by the Ministry of Land, Infrastructure, and Tourism of Japan.

As for fishing vessels using hydrogen, Ezoe and Takahashi (2008) introduced a small boat in which hydrogen gas was injected into the engine. This work also indicated the possibility of introducing hydrogen engines to small fishing boats. A fishing boat coupled with a system of engineering presented a trial opportunity to develop specifications for hydrogen fuel cells in coastal fishing vessels in 2010. However, these trials, especially for fishing vessels, were performed prematurely to the creation of the actual designs for shipbuilding because there were no guidelines and regulations at the time. Consequently, this did not lead to a practical use of hydrogen fuel cells for fishing vessels.

Toward the application of fuel cells in fishing vessels, automobile fuel cell stacks were adapted to fishing vessels with considerations given to future costs and ease of maintenance. The automobile industry is bigger than the marine industry, so equipment can be mass-produced in comparison to equipment developed specifically for marine use. The advanced technology required for the manufacture of such fuel cells is expensive if the products are manufactured solely for the marine industry. As for maintenance, fuel cell specialists are required to be located in fishing port areas, similar to the infrastructure and facilities required for shipbuilding and ship repair. Many services are already available for automobiles, and thus it would be efficient to borrow from the facilities and repair infrastructure currently available through the automobile industry to efficiently maintain fuel cell ships.

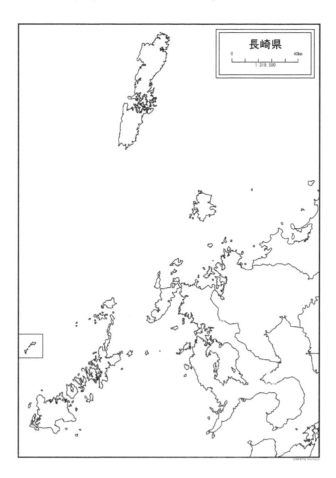

FIGURE 13.1   Goto Islands and Nagasaki Prefecture.

Recently, the hydrogen fuel cell boat safety guidelines for Japanese boats was established in 2018, and the need for utilization of renewable energy and hydrogen energy was accelerated. Scientists at the National Research Institute of Fisheries Engineering promoted a project for fisheries of Goto Islands with Goto City and Nagasaki Prefecture since 2014. This project was an attempt to develop a hydrogen fuel cell/battery fishing vessel using marine renewable energy. A chartered fuel cell/ battery boat launched in Goto City in 2015 was used to investigate the performance of a fuel cell-powered vessel in comparison to a diesel-powered vessel before designing the fuel cell-powered fishing vessel. We compared the propulsion, maneuverability, vibration, and underwater noise with data from the same hull type with a diesel-powered vessel and confirmed the efficiency of fuel cell/ battery boat (Takahashi et al. 2019). In this study, a trial design of a hydrogen fuel cell/battery-powered fishing vessel using an existing vessel was developed based on the 2018 hydrogen fuel cell ship safety guidelines. Development included considerations of the vessel's practical use in fisheries and usage of fuel cell stacks available for automobiles.

MATERIALS AND METHODS

TARGET VESSEL

There are approximately 80,000 coastal powered fishing vessels and 1,550 offshore fishing vessels in Japan. Most coastal fishing vessels make one-day fishing trips, and the size is just below 24 m in length and 20 gross tonnage (GT) in Japanese domestic GT. Offshore fishing vessels make trips

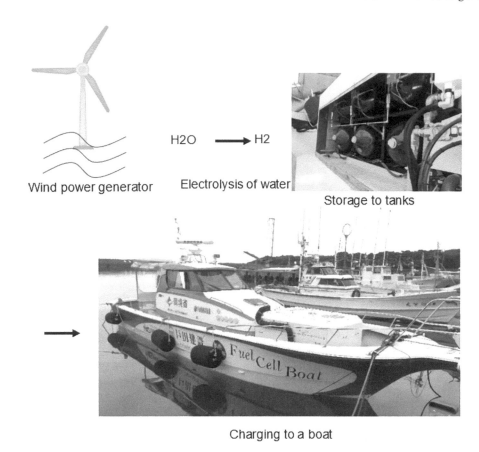

Wind power generator   Electrolysis of water

Storage to tanks

Charging to a boat

FIGURE 13.2   A schematic diagram of fuel cell delivery to a vessel (above) and (below) an actual boat using hydrogen derived from wind power in Goto City, Nagasaki Prefecture in 2015.

from several days to several years, depending on the type of fishing. The size of most of these ships is over 24 m in length overall. The engine output is regulated by the local government for coastal fishing vessels and by the central government for offshore fishing vessels.

As the first step for adapting hydrogen energy to a fishing vessel, we selected a bluefin tuna farming workboat (Figure 13.3) in Goto City, Nagasaki Prefecture. Bluefin tuna farming is a major industry in the Goto Islands. The Goto Islands are located in the west of Nagasaki Prefecture and have embayments suitable for fish farming. Boats can navigate to the calm sea and over short distances within 10 miles (Length of overall, LOA × Breadth, B × Depth, D = 25.01 × 5.48 × 1.82 m, Output 423 kW diesel). A farming workboat is registered as a fishing vessel in Japan. This vessel has a feeding machine powered by an air compressor, two cranes for capturing tuna, a net, and five capstans. A daily fishing trip is approximately 8 hours long. Figure 13.4 shows the main tasks in the daily work of these boats. Before departure, 10–15 tons of wild seed-stock are loaded onboard. After arriving at the tuna farming area, the workboat is moored with a culturing raft using capstans and the feed is thrown to the tuna by a feeding machine with an air compressor. Feeding is repeated several times using a culturing cage. Cranes are used to load tuna onboard.

FIGURE 13.3 Existing diesel-powered vessel for tuna farming that served as a reference vessel to design the fuel cell-powered vessel.

Loading wild-seed stock →

Tuna farming area

→

Feeding tuna

Capturing tuna

FIGURE 13.4 Tuna farming activities involving a typical workboat.

RESEARCH ON AN EXISTING VESSEL

Before the design of a hydrogen fuel cell/battery fishing vessel, we investigated the general arrangement, machinery, engine output, fuel oil consumption, and voyage trajectories of a target workboat in the field. To measure engine output (kW) and energy consumption (kWh) of an existing vessel, we used a satellite compass (Furuno Electric Co., Ltd., SC-30, IF-NMEASC) and a fuel flow meter (Hitachi Automotive Systems & Nagano, Ltd., AFOU631-74P) (Figure 13.5) on a target vessel. Flow meters were set into the fuel oil pipes in a machinery room, and data were collected by a data

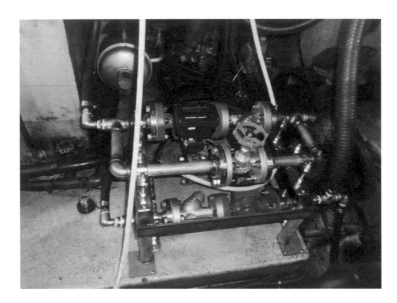

FIGURE 13.5   Flow meter installed into oil pipes in the machinery room of the fishing vessel.

logger (Kaiyo Denshi Co., Ltd., K-16035) over a year. The energy consumption was estimated with a constant fuel consumption rate of 0.27 L/kWh. The fuel consumption rate varied because of engine output load, but in this study, the average value was adopted.

Engine output was measured by speed trials near a tuna farming area (Figure 13.6). The speed was set to six points, and the energy output was determined using the slow engine speed around tuna farming areas, normal operation speed from port to the tuna farming area, and maximum speed. The average in 1 mile with return ways was adopted. Wind, current, and wave collection were not conducted because of the calm sea in the embayment. Engine output and energy consumption were used to decide motor and fuel cell stack power and battery and hydrogen energy capacities.

To investigate the principal particulars and current conditions of ship hull, the ship hull form was measured by laser distance and angle meter (Figure 13.7) during the period of maintenance on the dock. After obtaining data on a ship hull with x, y, and z coordinates, curves to obtain the ship hull form was constructed by non-uniform basic spline (NURBS), and lines and principal particulars were analyzed.

POWER UNIT

The power unit system consists of fuel cell stacks, lithium-ion batteries, hydrogen tanks, and a power control unit, as shown in Figure 13.8. The converter between the fuel cell and batteries is for boosting voltage. An inverter transduces the DC of the battery to AC to drive the motor. In this study, batteries were charged by all-electric energy generated by fuel stacks, and these batteries power the motor. The fuel cell stack power is not directly used to drive the motor. The efficiency for generating electric energy of an FC stack was assumed to be 50% at 25% FC stack output load in this study and was kept constant during FC stack moving. FC stack efficiency is shown in Figure 13.9. The most efficient FC stack output load is from 20% to 30%. Beyond 30%, almost all energy is consumed by the air compressor for cooling and is emitted as heat. A Toyota FC stack with 114 kW power was adopted in this study.

DESIGN OF GENERAL ARRANGEMENT

The general arrangement for the compartment and equipment is basically the same as that of the existing target vessel except the machinery. For general usage, the arrangement was to design the boat for main tasks of farming workboats (e.g., feeding and capturing tuna except net hauling of

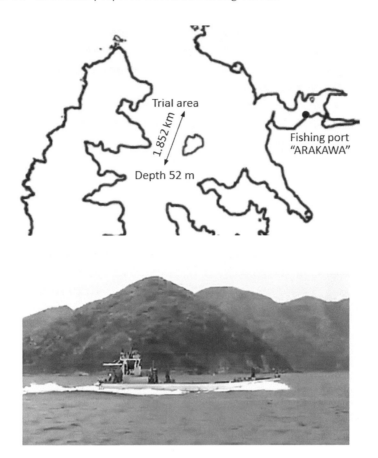

FIGURE 13.6 Sea trial area (above) and (below) vessel operating during sea trials when fully loaded.

FIGURE 13.7 Laser used to measure vessel dimensions and angles for describing hull shape.

farming cage). The front crane used for net hauling was removed. A power unit was installed as per the 2018 hydrogen fuel cell boat safety guideline by the MLIT, Japan. This guideline applies to small boats below 20 Japanese gross tons, and in the case of boats with more than 20 gross tons, it applies for sports and recreation boats below 24 m in length of the overall vessel. The power unit is a hydrogen and battery hybrid system, and the fuel cell stack is a polymer electrolyte fuel cell (PEFC). The pressure of the hydrogen tank is below 70 MPa. Hydrogen fuel, except for hydrogen reformed

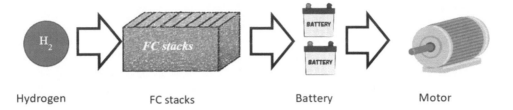

Hydrogen   FC stacks   Battery   Motor

FIGURE 13.8 Components of power unit.

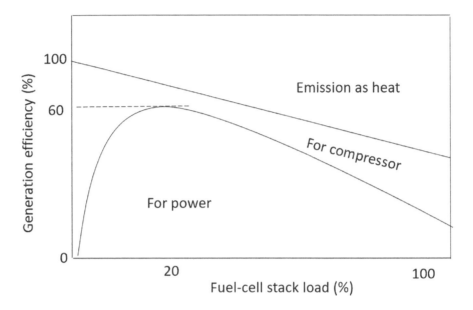

FIGURE 13.9 Relationship between fuel cell stack load and efficiency of electric generation.

natural gas, is pure hydrogen. The compartment of the fuel cell stack and hydrogen tanks are closed to each other. The navigation area is in a calm sea and in limited coastal areas. The guidelines indicate the approval, requirements, arrangement, and facilities for hydrogen fuel cell/battery boats.

For weight and trim conditions, the weight and energy power capacity of the batteries were determined first, because batteries have the largest weight of power units. An existing target vessel has a diesel engine and fuel oil tanks and attached gear and pipes. As alternatives to this machinery, a power unit was introduced. As the weight of a diesel engine with gears and fuel oil of an existing target vessel was approximately 3,000 kg and the weight of the motor was estimated to be 1,000 kg, the weight of batteries should be within 2,000 kg. Other parts of the power unit were estimated to be approximately 1,000 kg for hydrogen tanks and 200–300 kg for fuel stacks and power control units. The parts of the power unit except batteries and motor were added weight (approx. 1,500 kg) compared to an existing vessel. For space for the power unit, the batteries and the motor were arranged in a machinery room below the upper deck, and the hydrogen tank and fuel cell stacks were arranged above the upper deck close to each other. The hydrogen tank pressure was set at 35 MPa and 70 MPa. The size of the 35-MPa tank can vary, and the 70-MPa tank is adopted for automobiles. The tank energy capacity was determined and showed that the weight was approximately 1,000 kg.

ENERGY FLOW ESTIMATION USING WIND POWER

To estimate the potential of renewable energy and hydrogen storing and transportation, the energy flow shown in Figure 13.10 revised the energy flow of Mitsusima (2013). This flow consists of two

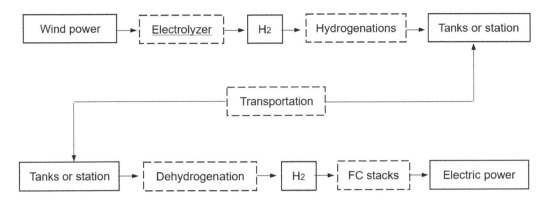

FIGURE 13.10   Energy supply chain from wind power to fuel cell fishing vessel.

processes. The first process is to produce renewable energy and store hydrogen in tanks or hydrogen stations in the field. After that, hydrogen is transported to different areas with tanks by trucks or ships. The second process is to consume hydrogen. In this study, for marine renewable energy, an offshore wind-power generator that can contribute to fisheries as a fishing reef was selected. Using the energy flow diagram, the relationship between the wind-power generator and hydrogen fuel cell/battery fishing vessel was obtained. The utilization factor of the facility and the availability factor of the facility was assumed to be 15% and 85%, respectively. The utilization factor of the facility is the factor that a wind-power generator is assumed to drive at maximum output load in a year. The availability factor of the facility is the factor that a wind-power generator is assumed to drive, except for the maintenance period.

RESULTS

GENERAL ARRANGEMENT, ENGINE OUTPUT, AND ENERGY CONSUMPTION OF THE EXISTING VESSEL

Principal particulars are shown in Table 13.1, and capacities of the compartment are shown in Table 13.2. The general arrangement and hull form of an existing vessel are indicated in Figure 13.11 and Figure 13.12. This vessel has a 76-ton displacement, 49-m³ fish cargo capacity for wild seed-stock, and 27-m³ machinery room capacity including fuel oil tank capacity.

TABLE 13.1
Principal Particulars of the Target Vessel, a 19 Japanese Gross Tons Tuna Farming Workboat

Principal particulars	Unit	Value
Draft (m)	m	1.35
Displacement (ton)	ton	76.166
Surface area (m²)	m²	156.282
Block coefficient, Cb	–	0.341
Prismatic coefficient, Cp	–	0.771
Center of buoyancy, lcb (aft, +)	m	1.344
Lcf (aft, +)	m	1.633
Propeller diameter, Dp	m	1.115
Propeller pitch, P	m	0.77
Num. of propeller blades	Blades	4
Ratio of expand area	–	0.60

TABLE 13.2
Capacity of Compartments

Compartment	Unit	Value
Fresh water tank	m³	5.1
Steering gear room	m³	5.22
Store No. 1	m³	6.52
Fuel oil tank	m³	2.27
Machinery room	m³	24.4
Pump room	m³	11.08
Fish hold No. 1	m³	25.42
Fish hold No. 2	m³	23.96
Store No. 2	m³	8.48
Store No. 3	m³	6.68

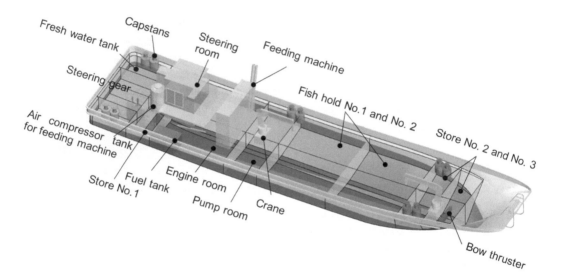

FIGURE 13.11 General configuration of an existing, diesel-powered tuna farm workboat.

The speed power curve and fuel consumption in speed trials under full load conditions are shown in Figure 13.13 and Figure 13.14. The engine output at normal navigation speed, 10 knots, is approximately 250 kW.

A sample of ship speed, engine speed, and output load variation from departure port to arrival port in a day trip on December 20, 2016, are shown in Figure 13.15. As for daily working, this workboat departed port for the tuna farming area at normal operational speed, 10 knots, in the morning. After arriving at the farming area, the workboat was moored using the farming raft, and fed the tuna. The main engine was driven at about 50 kW at an average for the feeding machine with an air compressor. After feeding, the workboat moved to the next farming raft and the same operation was repeated several times. At lunch time, the workboat come back to port and the engine was stopped. The same operations were repeated after lunch, and the boat came back to port in the evening. The total energy consumption was approximately 435 kWh.

The frequency of energy consumption in the daily trip during 2017 is shown in Figure 13.16. The peak frequency in a day trip was from 300 to 400 kWh.

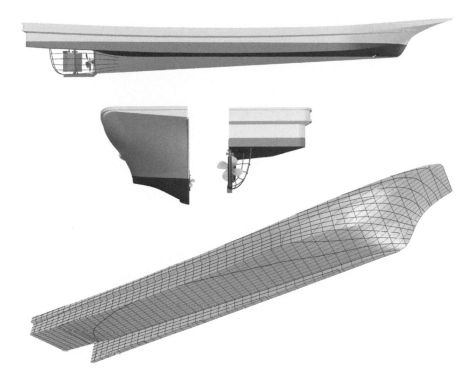

FIGURE 13.12 Ship hull form.

Power Unit

We planned to introduce batteries, fuel cell stacks, hydrogen tanks, and a motor, considering the weight and trim conditions of the ship and to satisfy the motor output and energy consumption for several trips. Table 13.3 shows the batteries and the motor specification, and Table 13.4 shows the hydrogen tank specifications at different tank pressures. The requirement of hydrogen was 24 kg H_2 for a trip of 435 kWh considering an energy density of 33 kWh/kg-H_2 and an FC stack generating efficiency of 50%.

An example of the energy consumption simulation is shown in Figure 13.17. This simulation was satisfied with the simulation of Figure 13.15. The nominal operational speed was assumed to be 10 knots, and the output of the motor was 200–250 kW and 50 kW while feeding and capturing at 0 knots, respectively. The electric energy charged by fuel cell stacks had remained in the batteries from departure port to arrival port.

General Arrangement

Three general arrangements at different tank pressures based on the safety guidelines are shown in Figure 13.18 and Figure 13.19. The detailed machinery arrangement for the power unit is shown in Figure 13.20. The hydrogen tanks were set on deck, and the fuel cell stacks were installed in a compartment close to the hydrogen tanks. Batteries, a motor, and a power unit control were arranged in the machinery room. These arrangements enable daily work onboard and the vessel has nearly the same fish cargo capacity as that of an existing vessel. The front crane, weighing approximately 1,000 kg and used for net-hauling, was removed, and the total displacement was nearly the same as an existing vessel, as shown Table 13.5. The center of gravity of the ship was, however, moved to the aft part, approximately 10,000 kg-m compared to an existing vessel. The farming workboat usually operates by filling the fish cargo hold with water. The trim balance can be easily taken as a

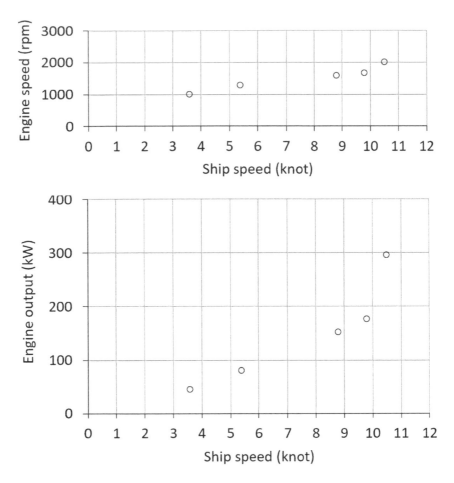

FIGURE 13.13   Result of speed trials: relationship between engine speed (above) and engine output (below) according to ship speed in knots for the diesel-powered reference vessel.

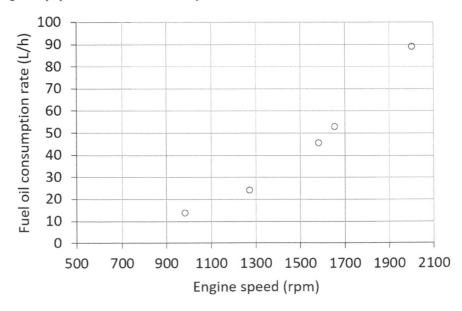

FIGURE 13.14   Rate of fuel oil consumption according to speed during speed trials.

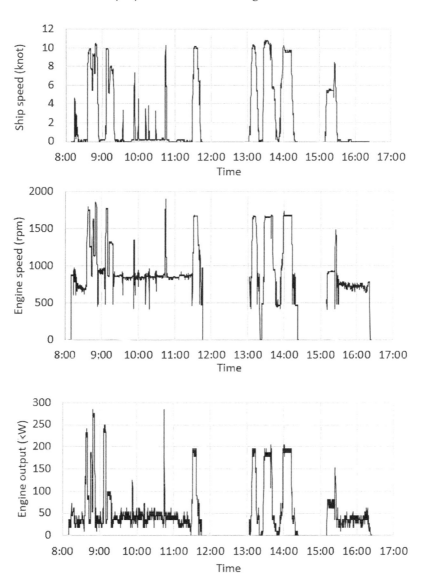

FIGURE 13.15 An example of results of vessel speed (above), engine speed (middle), and engine output (below) according to time of vessel operation for a diesel vessel.

100,000 kg-m moment adjustment capacity of fish cargo. In each case, arrangements for the target vessel were possible.

ENERGY FLOW ESTIMATION USING WIND POWER

The estimation of energy flow using Figure 13.10 from the wind-power generator to the hydrogen fuel cell/battery tuna farming workboat is shown in Table 13.6. The electric power of the wind-power generator was assumed to be 2 MW (2 MW × 1) in Case 1 of the actual situation offshore in Goto, and 22 MW (2.1 MW × 8, 5.2 MW × 1) in Case 2 of a future plan offshore in Goto in this simulation. The electric energy generation was, for example, calculated using 2 MW × 24 h × 365 days × 15% × 85% in Case 1. The total electric energy acquired from the wind-power generator was 2,234 MWh in a year. The final acquired electric power through hydrogen storing and transportation, considering efficiency of facilities, transportation, and generator was 726 MWh. Regarding the

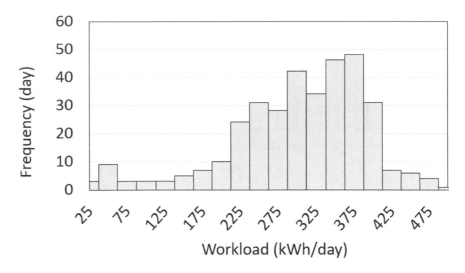

FIGURE 13.16 Frequency of energy consumption according to vessel workload.

potential workability of the hydrogen fuel cell fishing vessel, the wind power energy potential was equivalent to the energy consumption of five vessels in a year. In Case 2, the utilization factor of the facility was assumed to be 25% in this study, as a larger wind-power generator was introduced, and the energy potential was equivalent to the energy consumption of approximately 100 vessels.

DISCUSSIONS

The author investigated the possibility of a design for a hydrogen fuel/battery hybrid-powered fishing vessel based on the safety guidelines and usage of fuel cell stacks for automobiles. As the guidelines were formulated, the basic design of such boats has become applicable in Japan. A design was presented considering the practical usage of an existing vessel. Most of the past trials to introduce hydrogen fuel cell/battery boats were concept level, or for passenger ships, and there is no design for fishing vessels in practical use.

The analysis of the target vessel reveals the detailed arrangements, hull form, engine output, and energy. The number of workboats from 10 to 20 gross tons engaged in aquaculture in Japan is 1,557 (Census of Fisheries 2013). The results of the analysis can be applicable to estimate possibilities for these vessels to be used as hydrogen fuel cell/battery-powered workboats.

TABLE 13.3
Specifications for Motor, FC Stack, and Lithium-Ion Battery

Item	Value
Motor output and efficiency	350 kW, 95%
Motor weight and size	1,000 kg, 1,000 mm × 500 mm × 500 mm
FC stack maximum output	114 kW (Toyota Motor Corp.)
FC stack sets and total weight and size	3 sets, approximately 500 kg, approximately 1,800 mm × 1,000 mm × 1,000 mm for 4 sets
Lithium-ion battery energy capacity	150 kWh, 1,800 kg
Lithium-ion battery weight and size	Approximately 1,800 mm × 500 mm × 500 mm × 2 sets

TABLE 13.4

Specifications for Two Different Kinds of Tank Pressure of Hydrogen Tanks

	Case 1	Case 2
Compressed hydrogen tank pressure	35 MPa	70 MPa
Tank size and weight	2,342 mm, φ509 mm, 350 L, 112 kg (Hexagon Composites ASA)	1,900 mm, φ450 mm, 190 L, 140 kg (Tomoe Shokai Co., Ltd)
Tank hydrogen capacity	8.4 kg H_2	7.625 kg H_2
Tank sets	8 bottles	8 bottles
Total tank weight	896 kg	1,120 kg
Total tank space on deck	Approximately 9.5 m³	Approximately 6.8 m³
Total hydrogen capacity	67.2 kg H_2	61 kg H_2
Workability	2.8 days	2.5 days

The arrangement of hydrogen tanks and fuel cell stacks is better with a aft steering room, considering workability. The deck space with an aft steering room is sufficient for the arrangement of the hydrogen tanks and fuel cell stacks. The existing target vessel has a 423 kW diesel engine. The design engine load, considering a nominal operational load of 75% and 15% sea margin is 276 kW (= 423 × 0.75/1.15). The actual engine load was, however, from 200–250 kW (Figure 13.15). This

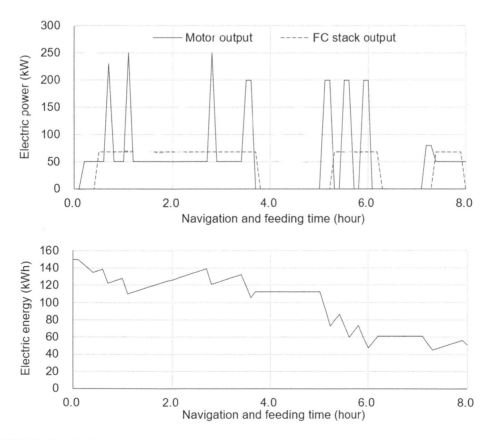

FIGURE 13.17 Simulation of motor and fuel cell stack output and energy consumption on a fishing trip.

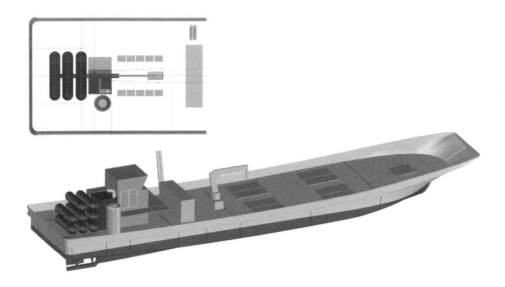

FIGURE 13.18   General arrangement of fuel cell components on a vessel with 35-MPa hydrogen tanks.

means that the captain of the ship reduces ship speed to save fuel, because the fuel oil consumption rate (L/h) will be 1.5 times higher if the engine load becomes 276 kW (Figure 13.14 and Figure 13.15). The energy capacity for a hydrogen fuel cell/battery workboat is suited to the level of energy-saving navigation in this study.

The fishing vessel consumes considerable energy during navigation and fishing. The hull space seemed not to be enough to install hydrogen tanks and the fuel cell and battery unit. Although the energy generated by a hydrogen fuel cell was thought to generate less power compared to diesel, the design indicated the possibility of hydrogen fuel cell/battery-powered fishing vessels.

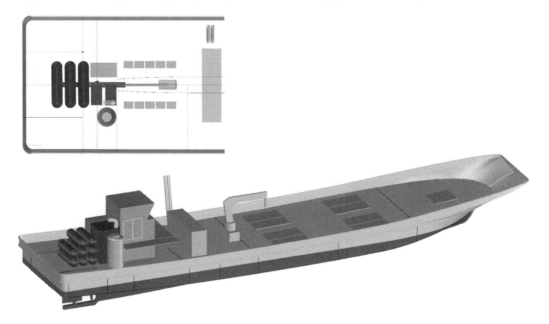

FIGURE 13.19   General arrangement of fuel cell components on a vessel with 70-MPa hydrogen tanks that are smaller tanks than 35-MPa hydrogen tanks.

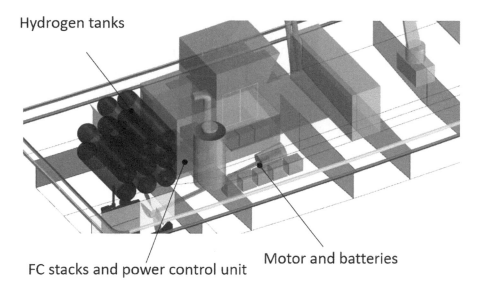

Hydrogen tanks

FC stacks and power control unit **Motor and batteries**

FIGURE 13.20 Arrangement of hydrogen fuel cell components, according to safety guidelines for a workboat.

TABLE 13.5

Weight and Trim Comparison between an Existing Vessel and Hydrogen Fuel Cell Vessel

Existing vessel/Hydrogen fuel cell vessel	Weight	Center of gravity (aft; +)
Main engine and fuel tanks/motor and batteries	3,000 kg/3,000 kg	10,000 kg-m/10,000 kg-m
Front crane/Hydrogen tanks and FC stacks	1,000 kg/1,000 kg	−10,000 kg-m/10,000 kg-m

Both 35-MPa tanks and 70-MPa tanks were used on deck, and the capacities were equivalent to a fishing trip lasting several days. The efficiency of the fuel cell stack electric power generation was 50% to 60% at 20% to 30% FC output load. Outside this fuel cell output load, most energy was consumed by the air compressor for cooling the FC stack and was emitted as heat. To support the constant requirement of 50 kW, an FC stack with 100 kW needs 3–4 sets. A battery capacity of 150 kWh is enough to maintain energy for the whole day.

As this guideline can be applied to small boats navigating in a calm sea and coastal areas, hydrogen fuel cells and battery power units can be applied to fishing vessels for aquaculture. There are 1,557 fishing vessels or workboats from 10 to 20 gross tons in the Japanese aquaculture industry. There are 2,790 fishing ports in Japan and 475 fishing ports on the islands (Japan Fisheries Agency 2017). Though the design is limited in small boats within the guidelines, the potential for storing hydrogen energy is good not only because of the profitability of the hydrogen supply, but also as an energy supply in case of natural disasters.

A wind-power generator was installed offshore near Goto Island and will be producing approximately ten times the electric power in the future. Wind-power generation has the potential to be applied to fishing reefs. As fisheries are the main industry in remote islands, the utilization of wind power for fish farming and workboats can be considered.

FUTURE WORK

We are developing better-suited machinery and deck arrangements considering profitability and technological progress.

TABLE 13.6

Electric Power Flow Simulation from Wind Power to a Hydrogen Fuel Cell Tuna Farming Workboat

Item	Unit	Case 1	Case 2
Electric power of a wind-power generator	MW	2	22
Utilization factor of facility	%	15	25
Availability factor of facility	%	85	85
Electric energy generation in a year	MWh/year	2,234	40,953
Electrolyzer efficiency	%	80	80
H_2	MWh/year	1,787	32,762
Hydrogenation efficiency	%	90	90
Energy carrier	MWh/year	1,608	29,486
Transportation efficiency	%	95	95
Energy carrier	MWh/year	1,528	28,012
Dehydrogenation efficiency	%	95	95
H2	MWh/year	1,452	26,611
	kg-H2/year	43,986	806,402
Generator (Fuel cell stack) efficiency	%	50	50
Electric power	MWh/year	726	13,306
Energy consumption of a hydrogen fuel cell fishing vessel in a trip	kWh/day	435	435
Days at sea	Days	310	310
Energy consumption in a year	MWh/year	135	135
Availability of the number of vessels	Vessels	5	98

ACKNOWLEDGMENTS

I am grateful to Nagasaki Prefecture, Goto City, Tuna Dream, Toyota and Toda Corporation for data measurement and discussion. I also thank the Fishing Boat and System Engineering Association of Japan, National Marine Research Institute and National Fisheries University for professional advice. Additionally, I also thank the project team members, director Yoshimi Takao, senior researcher Hiroyasu Mizoguchi, researcher Ryuzo Takahashi and researcher Katsuo Hasegawa of FRA. This research was achieved with their assistance.

REFERENCES

Barrett, S. 2018. Toyota begins sales of Sora fuel cell bus. *Fuel Cells Bull.* 2018:2–3.

Census of Fisheries. 2013. Ministry of Agriculture, Forestry and Fisheries, Japan.

Ezoe, S., and K. Takahashi. 2008. Trial production of outboard type hydrogen-gas combustion engine for small fishing boat. *Japan Soc. Mech. Eng.* 088–1:317–318.

Hirata, K., K. Kiyokawa, T. Hainiwa, T. Hiraiwa, and F Yukizane. 2019. Our commitment to developing hydrogen fuel cell ship. *Mar. Eng.* 54:156–9.

Hirata, K., and K. Miyazaki. 2017. Design and manufacturing of model hybrid boat with hydrogen fuel cell and lithium-ion battery. *Translog2017* (in Japanese), No. 17.

Japan Fisheries Agency. 2017. List of the number of fishing ports (in Japanese). Available at: https://www.jfa.maff.go.jp/j/kikaku/wpaper/h29_h/trend/1/sankou_5.html#a01

Kifune, H., M. Satou, and T. Ode. 2016. A study on battery system design for battery ships conforming to CHAdeMO. *J. Japan Inst. Mar. Eng. Eng.* 51:879–86.

Maekawa, J., B.H. Hai, S. Shinkuma, and K. Shimada, K. 2018. The effect of renewable energy generation on the electric power spot price of the Japan electric power exchange. *Energies* 11:2215.

Mitsushima, S. 2013. Energy storage and transportation technology for huge scale renewable energies. *GS Yuasa Technical Report (in Japanese)* 10:1–6.

Obara, S. 2019. Study on improvement of the utilization factor of a transmission network and increase in the amount of renewable energy introduction by hydrogen energy careers (Model example of the North-Hokkaido Wakkanai area). *Trans. JSME (in Japanese)* 85:18–00141.

Ode, T., and E. Shimizu. 2019. Current status of and trends for battery-powered ships and fuel cell ships in Japan - Research and Development at Tokyo University of Marine Science and Technology. *Mar. Eng. (in Japanese)* 54:598–605.

Takahashi, R., J. Miyoshi, H. Mizoguchi and D. Terada. 2019. Comparison of underwater cruising noise in fuel-cell fishing vessel, same-hull-form diesel vessel, and aquaculture working vessel. *Transactions of Navigation* 4:29–38.

Toda Corporation. 2015. *Fuel Cell-Powered Marine Craft*. Ministry of Environment, Japan.

Utsunomiya, T., I. Sato, O. Kobayashi, T. Shiraishi, and T. Harada. 2015. Design and installation of a hybrid-spar floating wind turbine platform. *Proc. Int. Conf. Offshore Mech. Arct. Eng. - OMAE* 9:V009T09A063.

van Biert, L., M. Godjevac, K. Visser, and P.V. Aravind. 2016. A review of fuel cell systems for maritime applications. *J. Power Sources* 327:345–64.

14 Summary
The Future of Fisheries Engineering

Stephen A. Bortone and Shinya Otake

Well, not a summary, *per se*, as each of the chapters offered an Abstract and the salient points of each chapter were presented in the Introduction (Chapter 1). More to the point, this will be a brief summary of the future of Fisheries Engineering as a discipline in fisheries science on points derived from the International Conference on Fisheries Engineering (ICFE 2019). Below are highlighted the anticipated key directions for the future of Fisheries Engineering. Practically, however, one might assume that future trends in Fisheries Engineering are more probably going to be continuations of the current trends, but driven by new technology and advances now available owing to more cooperative, multidisciplinary investigations.

As indicated in the Introduction, the chief goal of Fisheries Engineering is "to sustain fisheries and its associated human socio-economic communities through the application and development of information and technology." While sustaining fisheries for the future is paramount, one must be cognizant of the effect that the potential of human-induced climate change may have on the environment and how fisheries respond to these changes. In fact, how Fisheries Engineering responds to the future challenges brought on by climate change will likely highlight most future efforts in this field.

It is unlikely that this overall goal of Fisheries Engineering will change appreciably in the near future. Kimura (Chapter 2 in this volume), in his keynote review, noted several areas that are currently receiving notable attention in Fisheries Engineering research. These areas include: artificial reefs, advances in underwater surveys, the application of computational fluid dynamics technology, and (as implied by Kimura) changes in fishing laws. Notably, Kimura indicated the tremendous potential that the use of artificial intelligence in both fisheries and aquaculture will have on the future of Fisheries Engineering, as well as the improvement of DNA analysis in underwater environmental analysis of communities. These areas will continue to receive the attention of the Fisheries Engineering community of scientists, managers, and user-groups.

One purpose of the conference upon which this volume is based was to clarify the global use of artificial reefs with an emphasis on differences according to regions. Lima and Zalmon (Chapter 3) stated that trends in artificial reef research would continue with an emphasis on gleaning a better understanding of settlement mechanisms, fishery production, and environmental impacts of human activities. They indicated that progress in the use of artificial reefs would likely come from areas involving development of new reef materials and methods of data analysis. Both of these benefit from a future with more long-term, broad geographical studies that include socio-economic features of artificial-reef associated fisheries.

Pioch et al. (Chapter 4) saw a future of artificial reefs in France more directed toward restoring biodiversity, especially in view of the portended influence of climate change and other human influences. As a model and example for other nations, France is pursing investigations that make use of artificial reefs to mitigate human impacts to aquatic ecosystems. Moreover, the authors envision a future where artificial reef deployments can benefit fisheries by combining the positive features of the natural environment with human activities.

Lee et al. (Chapter 5) realize that many questions remain unanswered in Korea's pursuit to make better use of artificial reefs in the future sustainability of its fishery resources. These questions, relating to proof of the positive relationship between fisheries resource features such as biodiversity and production, will remain at the forefront of future research investigations.

In Southeast Asia, artificial reefs are often constructed using locally grown vegetation. These reefs provide habitats and are utilized to facilitate fish harvests that subsequently help build a local fishing economy. While still in various stages of development and improvement, the technology involved is necessarily fairly primitive. Jani (Chapter 6) foresees a future for Fisheries Engineering, especially as it pertains to artisanal artificial reefs, that includes involvement of a broader range of stakeholders to achieve fishery management solutions. Thus, merging the knowledge base known to users with the information and analytical expertise of scientists and resource managers will produce the best solutions.

Otake (Chapter 7) takes a critical, long-view approach to long-term solutions toward improving fisheries by suggesting large-scale artificial reef deployments to create extensive fishing grounds. While currently these grounds do not dominate the catch in any fishery, expectations are that these artificial reef-assisted fishing grounds will comprise a larger portion of the total catch for some fisheries in the future.

Inoue et al. (Chapter 8) offer an excellent example of how future evaluations of fisheries data may take place. By challenging the assumptions inherent in past analyses, researchers can more clearly discern the underlying factors responsible for observed responses. Here, Inoue et al. challenge how CPUE has traditionally been examined and offer a reasonable, and intuitive, way to elucidate the underlying forcing factors that abound in complex situations. Fisheries Engineering and other complex environmental sciences can greatly benefit from such examinations using this fresh perspective.

Seagrass beds have and will continue to play an important role in the coastal environment, but sea-burning (isoyake) occurs frequently in Japan and in other temperate regions of the world. Hashimoto et al. (Chapter 9) introduce the countermeasure technologies that are being promoted in Japan. Hashimoto et al. offer new solutions to this perplexing problem that involve the decimated grassbed substrates that plague certain coastal areas. Their solution to incorporate artificial reefs into the solution demonstrates to future fisheries engineers that other advancing technologies can often be "borrowed" to offer alternative solutions.

Takahashi et al. (Chapter 10) expand investigations into the personal safety aspect of fisheries to make use of testing methods often used in modern consumer sciences to help give direction to solve a long-standing problem. Collectively, there is much to learn here. Safety issues associated with fishing will be addressed in the future. The approach will include not only confronting the direct solution of safety for fishermen through more compatible equipment, but indirectly through avoiding unsafe conditions through the development of robotics to replace the more dangerous activities by fishermen. In the future, we will move toward securing simple and safe fishing labor, and in the future, we will robotize it to help eliminate dangerous work and prevent a decline in the workforce due to population decline. Empirical, experiential data of real-world conditions can often enlighten problematic situations. Future examinations of other problems in fisheries may do well to follow this line of investigation, especially with regard to problems confronting fishermen and their safety.

The symposium on the last day of the conference included presentations and discussions relative to the significance of offshore wind turbines in Fisheries Engineering. It has become clear that there is potential habitat value in the structural, benthic component of these wind turbines. Although there are controversies in the deployment of these structures, with cooperation, the opposition from fishermen can be minimized. Presently, in Japan, fishermen have exercised their rights, and they are often seen as a symbol of opposition to construction of wind turbines. Naturally, wind power systems will increase their presence in the vast oceans in the future. It is expected that the fishing grounds will change drastically, and fishermen will be in opposition to them. From the viewpoint of Fisheries Engineering, it is necessary to investigate the factors that may create and improve fishing grounds, perhaps through the construction of these artificial reef-like structures.

Dr Nakata gave a deep and broad introduction to the formation of fishing grounds through the construction of wind-power generators. Nakata (Chapter 11) proposes that creating new habitats (e.g., similar to the habitat creation in artificial reef deployment) can take place in a sensible manner and with consideration given to fishermen. This practice is becoming even more common with the expansion of offshore renewable energy-related structural deployments. Being mindful of the harmonious compatibility of these structures with the natural environment will go far in efficiently achieving fishery objectives that involve habitat improvement and modification.

Pol and Ford (Chapter 12) introduced the idea that wind power facilities are improving the habitat effect that construction of these facilities provides in the United States, expanding on the idea of coupling Fisheries Engineering with efforts to expand the development of renewable energy. The problems related to the deployment of structures for renewable energy are similar to those confronted by those interested in deploying artificial reefs. Pol and Ford take this opportunity to demonstrate how efforts to facilitate the fishing community's acceptance and accommodation to these deployments can serve as models to facilitate marine deployments in other areas such as artificial reefs. Future efforts will likely continue to align renewable energy interests with those of the fishing community by eliciting mutual cooperation for mutual benefits.

Miyoshi (Chapter 13) offers a more directed view of Fisheries Engineering in the future by demonstrating the accommodation that fishers can make by adopting more energy-efficient propulsion for their vessels. This chapter clearly shows that Fisheries Engineering can make considerable contributions to the future direction of energy savings and, indirectly, help improve the environment.

As described above, this volume introduces many aspects of the latest technology in Fisheries Engineering and, consequently, could be used as a textbook or model to give direction to future efforts by engineers and researchers. While much of the technology proposed is relatively simple, combinations of technologies in concert with new perspectives can allow for the development of new ideas and innovative approaches to persistent problems.

No one of the above examples of future directions for Fisheries Engineering will dominate, but cumulatively, they will help achieve the ultimate goal of attaining sustainability of our fisheries and the associated socio-economic features of their communities. It is possible to show that the expansion of Fisheries Engineering will benefit by more broadly coupling past methods with modern advances in technology. Fisheries Engineering, as a discipline, is likely to gain most from the advent of new technologies and their implementation in innovative ways. These technologies include genetic engineering, new materials, facilitated information transfer, and refined statistical treatments of already existing data. Stay tuned for the accomplishments of our colleagues as they continue to address some of the most important environmental, cultural, and socio-economic issues of our time.

Index

T - #0531 - 071024 - C168 - 254/178/8 - PB - 9780367560669 - Gloss Lamination